Gebhardt
Rapid Prototyping

Andreas Gebhardt

Rapid Prototyping

HANSER
Hanser Publishers, Munich

Hanser Gardner Publications, Inc., Cincinnati

The Author: Prof. Dr.-Ing. Andreas Gebhardt
Aachen University of Applied Sciences, Aachen, Germany
General Manager CP – Centrum für Prototypenbau GmbH, Erkelenz/Düsseldorf, Germany

Distributed in the USA and in Canada by
Hanser Gardner Publications, Inc.
6915 Valley Avenue, Cincinnati, Ohio 45244-3029, USA
Fax: (513) 527-8801
Phone: (513) 527-8977 or 1-800-950-8977
Internet: http://www.hansergardner.com

Distributed in all other countries by
Carl Hanser Verlag
Postfach 86 04 20, 81631 München, Germany
Fax: +49 (89) 98 48 09
Internet: http://www.hanser.de

The use of general descriptive names, trademarks, etc., in this publication, even if the former are not especially identified, is not to be taken as a sign that such names, as understood by the Trade Marks and Merchandise Marks Act, may accordingly be used freely by anyone.

While the advice and information in this book are believed to be true and accurate at the date of going to press, neither the authors nor the editors nor the publisher can accept any legal responsibility for any errors or omissions that may be made. The publisher makes no warranty, express or implied, with respect to the material contained herein.

Library of Congress Cataloging-in-Publication Data

Gebhardt, Andreas.
Rapid prototyping / Andreas Gebhardt.-- 1st ed.
   p. cm.
ISBN 1-56990-281-X
1. Rapid prototyping. I. Title.
TS155.6.G43 2003
620'.0042--dc21
                        2003004412

Bibliografische Information Der Deutschen Bibliothek
Die Deutsche Bibliothek verzeichnet diese Publikation in der Deutschen Nationalbibliografie;
detaillierte bibliographische Daten sind im Internet über <http://dnb.ddb.de> abrufbar.

ISBN 3-446-21259-0

© Carl Hanser Verlag, Munich 2003
Production Management: Oswald Immel
Coverdesign: MCP • Susanne Kraus GbR, Holzkirchen, Germany
Typeset, printed and bound by Kösel, Kempten, Germany

A picture says more than 1000 words
– a model tells the whole story

Preface

"From an engineer's toy to the key to fast product development" proclaimed the caption in the Preface of the first German edition of this book.

At the end of 1995, about 6 years after the first stereolithography machines had been installed in Europe, rapid prototyping processes had developed into effective tools enabling product development to accelerate and improve. They had envolved from an occasionally used, technically fascinating but economically unattractive model-making procedure to a speed determining element in the product development chain.

Apart from stereolithography, other prototyping processes were also established. The range of materials that may be used was drastically increased. The physical properties of models could be considerably improved, and their characteristics approached those of the target material. Higher precision and faster machines resulted in better models at lower costs.

"From a tool for fast product development to a tool for fast product formation" best captions the development since 1995. The urgent desire for plastic or metal prototypes with series-identical standards was the driving force for the further development of already existing plastic processing methods, on which processes for metal, sand, ceramics, and other materials are based, and especially for processes that enable the production of molds and tools. These applications of rapid prototyping technology, collectively known as rapid tooling, and all processes that directly or indirectly facilitate the processing of metals therefore became the center of attention. They are of special interest as, being the interface between development and production, they effectively shorten the "step into the tool" which is time-consuming and expensive when traditional methods are used. This book therefore contains detailed sections on rapid tooling.

Processes that complement, limit, or to a certain degree, compete with rapid prototyping are discussed only insofar as necessary to make the rapid prototyping process understandable. This applies especially to methods of virtual reality, high-speed milling, and conventional processes. The same applies to the intensive discussion of CAD processes and reverse engineering. Bibliographical references enable the reader to further deepen his or her knowledge on these subjects.

We should be aware that this edition can present only a snapshot at this moment in time. The rapid prototyping market is still developing at lightning speed. The number of systems sold worldwide doubled between 1995 and 1997 and, after a poor 1998, is growing since at a nearly constant 200 systems per year. There is no visible end to this development, although slightly degressive growth rates seem to indicate that the scene is settling.

Rapid prototyping processes have established themselves as tools providing effective support in product formation processes. Their use therefore does not give a competitive edge. The reverse conclusion also applies: Not to use them means fighting with blunt weapons.

Andreas Gebhardt

Acknowledgements

The interdisciplinary character of Rapid Prototyping as well as the tremendous speed of its development make it almost impossible for a single person to present it completely, accurately and topically. Therefore, I am very thankful for the help provided by friends, colleagues, and companies.

Special thanks to my colleagues from the Center of Prototyping, Erkelenz, Germany, who provided the continuous contact to the 'shop floor' and the basis for the practical orientation of this book. Personal thanks go to *Christoph Schwarz* and *Michael Wolf*. *Besima Suemer* used numerous discussions to support the improvement of the book's structure and it's understandability.

Many additional details were obtained from the research work done at the Prototyping Center of the Aachen University of Applied Sciences.

Special thanks go to all who contributed to this book. Personal thanks to *Wolfgang Steinchen* [5.3], *Bernd Streich* [5.5], *Frank Petzoldt* [8.1], *Christian Wagner* [8.1], *Konrad Wissenbach*, *Andres Gasser*, *Eckhard Hoffmann* and *Wilhelm Meiners* [8.2] who provided major contributions to the chapters in brackets.

Personal thanks to *Mrs. Wolstenholme* for her help with the translation and *Carsten Tillmann* for double-checking all tables.

Contents

1 Product Development – Product Formation – Rapid Product Development

Rapid prototyping methods are developing quickly and steadily and have progressed from being tools for fast product development to becoming tools for fast product formation. The entire product formation process comprising product development, development of production facilities, and production itself is therefore considered here. Nevertheless, the early stages of product development have special significance for the success of a product in the market. This chapter discusses the reason for this, why certain methods help to optimize this process, the role models play, and the special role of rapid prototyping in this process.

1.1 New Demands – New Processes

1.1.1 Changing Conditions for Product Development – Critical Factors Affecting Success

Successful product development means developing a product of highest quality, at lowest costs, in the shortest time, in such a way that it can be produced quickly, safely, and at a reasonable price. The circumstances under which these aims are achieved today and will be in the future are constantly changing.

The ever increasing pressure to succeed combined with decreasing budgets will make it even more important in the future to recognize changing circumstances early and to develop suitable strategies to accelerate product generation processes.

Although it is hardly possible to generalize in respect to the changing requirements for products, it is possible to show trends. These are more pronounced for consumer goods and less visible for industrial goods, but they exist for both equally.

Assuming that customer desire and behavior define the requirements of a product, the following new or changed circumstances for product development result:

- *Non specific or fast changing customer desires*

  It becomes increasingly difficult to define what the customers want. For a long time now it has not been the properties of a product alone that have determined customer desires; trends, too, have influenced customer buying habits. Often, trends need to be detected by means of market surveys and products must be ready before the trend changes.

- *Increasing significance of styling*

  Providing a functional product with an attractive styling is increasingly important. In buying a certain product people tend to express their lifestyle.

- *Individualization of products*

  In spite of mass production the customer expects an individual product, custom made and distinctive if possible. Individual trimmings and technical details of cars demonstrate this impressively.

- *Environmental requirements*

  Environmental compatibility and the ability to recycle products and their packaging determine the buying behavior of customers as subsequent costs are dependent on them.

- *Decreasing lifetime of products*

  A drastic decline in the lifetime of all types of products is being witnessed. According to leading market research agencies the period of time over which a product can be placed profitably in the market has nearly halved over the last 20 years. Studies from various sources agree that this trend will continue at a rate of about 5% per year (Figure 1-1). The speed of change will largely depend on the line of business. Whilst the electrotechnical industry or suppliers to car manufacturers have, owing to their intensive use of new methods, already exhausted the potential to a great extent, more conservative lines of business such as the machine-tool manufacturing industry still have considerable potential [BUL97].

- *Decreasing prices*

  The price of a product is gaining increasing influence in the buying decision. Global markets and fast communication between continents enable a worldwide price comparison. Geographical niches can no longer be occupied, at least not for long.

In addition to these circumstances, characterized by individual market and customer desires, other changed circumstances become apparent in the form of changing regulations and standards. Technical progress, represented by new materials, development, and production methods that may be used in various ways, and different location criteria also influence the optimal design of a product.

- *Regulations and standards*

  Products are required to conform to national and international standards and regulations [CE-label (Conformité Européenne; *CE marking within the European Union),* electromagnetic compatibility (EMC), total quality management (TQM)]. Quality and environmental requirements not only influence styling, design, and production but also demand special solutions for numerous market segments thereby creating de facto admission barriers.

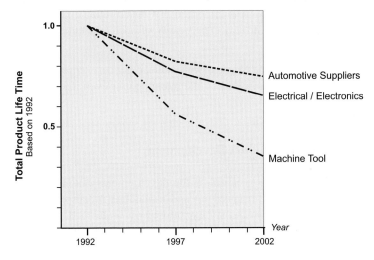

Figure 1-1   Change of product lifetime based on 1992 for various branches of industry over time

- *Interdisciplinary work*

  As ever more complex products and production processes appear, the theoretical possibilities for developing and producing a product increase exponentially. It is no longer possible to develop competitive products without the participation of experts from various disciplines in product development and production. As a result more than one team of experts from various trades need to be consulted when developing a product.

- *Spreading of resources*

  Siting criteria such as tax laws, wage levels, social security payments (additional labor costs), infrastructure, and raw materials have caused international companies especially to spread their resources and competence accordingly throughout the world.

- *Legal circumstances*

  An increasingly consumer-friendly jurisdiction is developing in all countries of the world which, in cases of product liability, establishes as a principle that the manufacturer is liable toward the customer irrespective of whether the manufacturer's direct negligence can be proven. These legal aspects cannot be discussed in this book in depth. Moreover the manufacturer is unable to influence them, but needs to be aware of them and to take them into account.

There is one area, however, that the manufacturer can and should actively influence: instructions for use and technical documentation. In view of the ever increasing lawsuits for damages, especially in the United States, this documentation issued by the manufacturer is gaining importance. The law requires not only that all effects the product could bring about be appropriately acknowledged (even if these are improbable), but especially that the instructions for use should be presented in such a way that the customer is able to understand them. Incomplete, unclear,- or even incorrect instructions will be viewed unfavorably for the manufacturer in a dispute. It is important that descriptions given in the instructions for use adopt the quality of a warranty. The quality of such warranties forms the basis of every liability lawsuit.

It follows that guarantees must be checked with painstaking care. Guarantees are subject to unlimited liability for a result promised by the manufacturer even if such a result may not normally be anticipated.

1.1.2   Consequences for Product Development

All areas of industry have tried to adjust to the developments outlined in the preceding. Methods have been introduced to speed up and rationalize processes. "Reserves" of all kinds have been reduced, thereby freeing capital tied up in personnel and material. Lean production was introduced in production plants and lean management pruned executive structures. "Just in time" delivery helped to reduce inventory. "Business reengineering" organized new internal structures and their interactions.

The general pressure to rationalize and the increased demands placed on the product reached through to product development some years ago.

Computer aided design (CAD) was introduced in the 1970s. At first, however, the various computer-aided processes did not bring about fundamental changes in product development strategies. In practice, the computer screen basically replaced the drawing board.

These changed circumstances have established new requirements in product generation and especially product development:

- Customer wishes need to be integrated faster into product development by market research. In addition, it is essential to reach agreement on the most important development aims as quickly as possible.

- The final styling focussed on a functional and attractive exterior must be decided upon at the earliest possible moment in the product development stage.

- The product needs to be designed in such a way that customized configurations reflecting the desires of various groups of customers can easily be accommodated. This should necessitate neither a new design nor the loss of the advantage of mass production.

- The product must retain its environmental acceptability during its production, its usage and afterwards.

- The product should sell in large quantities over a relatively short life cycle.

- Attractive profit margins are possible only so long as the product is unique on the market, that is, before any competitive products materialize. Profits for the financing of future product generations must be realized during this period.

- National and international standards and regulations must be considered parallel to product development. In certain cases, it may be necessary to consult external experts. Standards and regulations require adjustments or even new formulation (development accompanying standardization) to an increasing degree.

- Differing technologies, distributed competence and resources need to be considered. Teams from various branches speak different languages, literally and idiomatically. The importance of communication increases.

- Technical documentation and instructions for use must be finished and legally checked before the start of production.

To summarize, it follows that products of the future will be more complex. They will need to be developed faster, at lower costs, with better quality and for shorter product life cycles by larger interdisciplinary teams

1.1.3   Critical Factors for Success and Competitive Strategies

Critical factors for success are defined by Siegwart and Sieger [SIEGWART91] as measures by which single influences may be condensed, thereby enabling the measurement of the degree of success of a company.

The influences discussed in the preceding section and the resulting consequences for product development point to the following factors for success:

- Shortening of the product development time

- Reduction of costs

- Increase in flexibility (product and production)

- Improvement of quality

This list is not universally applicable but it does represent today's generally accepted consensus.

The single critical factors of success are not independent of each other in the mathematical sense; they represent values that, when weighted and interrelated in a strategy, enable conclusions to be drawn. The strategy of a company expresses how it considers its own interaction with its competitors. Today (it can be different tomorrow) the following strategies are pursued by successful enterprises (market leaders as well as "hidden champions"):

- Technology leadership (pioneer strategy)

- Cost leadership

- Differentiation

- Concentration

- Outpacing

The critical success factors of time and flexibility take priority in the pioneer strategy; quality and costs play secondary roles. In the cost leadership strategy all other critical success factors are subordinate to cost factors. The differentiation strategy aims for a unique product, a distinctive design perhaps. In this case cost and time tend to be subordinated to quality and flexibility. Concentration is not an independent strategy but rather the application of cost leadership, or differentiation for a chosen market segment. Outpacing means alternating between the two strategies of cost leadership and differentiation depending on the competitive situation. "Preventative outpacing" aims at immediate cost leadership to prevent low-price competitors from gaining an advantage a priori.

Similar to outpacing is the so-called market absorption strategy. A product priced as high as possible, as yet without competition owing to its early introduction to the market, is intensively marketed for a short time with high advertising expenditure. After the competition joins the market, there may no longer be any interest in the product, only a slow and extremely low-priced follow-up business remains. This is especially true for fashionable products with "cult status" such as the children's toy "Tamagochi." In this case all factors of competition play an equal role, but their weight varies from phase to phase.

This shows that combination strategies are used most often, with a variation in the composition of single factors that are emphasized to different degrees over time. It also becomes obvious that the critical success factor of time plays a pivotal role.

1.1.4   The Key Factor of Time

This section demonstrates that all critical factors for success of a product, but especially time and cost, can be condensed into one key element: the "time to market." "Time to market" means the time that elapses between the decision to develop and produce a certain product and its introduction into the market.

Analysis shows that cash savings are greatest when the time to market is minimized; the converse also applies. Figure 1-2 is based on a 1983 survey by McKinsey & Co. (USA) [REI83]. It deals with the influence of various success factors on company profit against the background of various strategies. The light-colored columns presented in countless articles on rapid prototyping in this or another form demonstrate the overproportional influence of "time to market" on company profits. The darkcolored columns show less growth, a smaller fall in prices, and a doubling of the life span of the product. Under these circumstances "time to market" loses its significance, while the importance of production costs increases. The business strategy therefore plays a central role when judging influences. The survey was conducted in 1983. Today, consumer goods with a life span of 5 years are considered durable products. As we have seen, product life spans have fallen dramatically and the trend is continuing. This makes clear that, for this reason alone, the "time to market" is becoming increasingly important for very nearly all products.

This example shows that if the budgeted development costs are exceeded by 50% the budgeted profit is reduced by only 3.5%, whereas an overshoot of 9% in the budgeted production costs results in a reduction of 22% in profit. Exceeding the planned development time by 6 months with a resulting late entry into the market, however, will reduce the planned profit by a third.

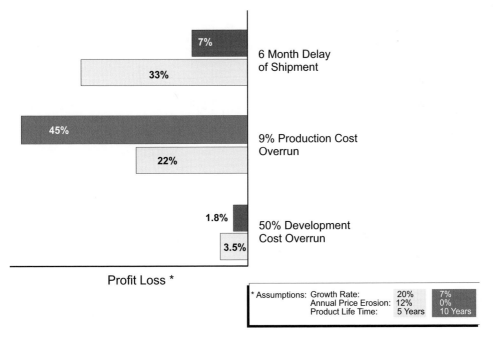

Figure 1-2 On the significance of "time to market": How missing the aim for various reasons influences the realizable profit

These results are confirmed by a survey made by Karjalainen and Tuomi [KAR98] concerning the development of car tires. With a product life span of 6 years and accounting for turnover degression caused by competition and declining attractivity with growing product age, they establish a 31% reduction of profit that they attribute solely to a delay of 6 months in entering the market.[1]

This shows clearly that "time to market" is the key factor for the success of a product and becomes the most important criterion for management.

This dominance of time over money, typical for today's products, has not only an absolute but also a relative dimension. It is not solely a question of making the right decisions within a short period of product development; of equal importance is making those decisions as early as possible. In addition it should be realized that although the accumulated expenditures for product development early in the process are still low, a large percentage of future costs is already determined over the course of the development. With the aid of the diagram

[1] The relationships are valid in analogue form for other products with different ratios of product life time to product development time, which lead to different results. There are no detailed surveys as yet.

published by Altmann it can be established that whereas after the completion of the draft phase only approx. 5% of the total costs have been incurred, 75% of the total costs have already been predetermined by the development (Figure 1-3) [ALT94].

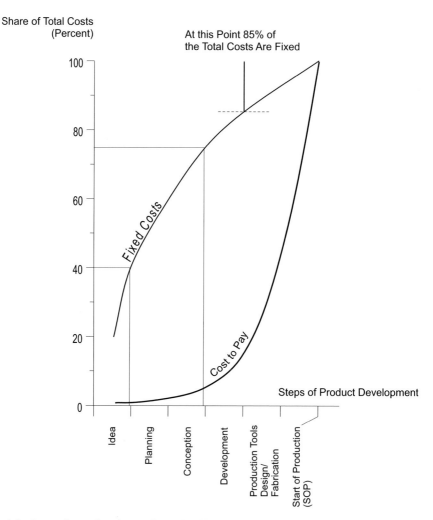

Figure 1-3 Comparison of total costs determined in the course of product development with the expenditures

The graph shows clearly that the relationships are even more dramatic at the commencement of product development: 40% of the total costs are already definitely fixed after the idea and draft phase, although at this point in time only negligible costs of 2% to 3% of the total costs have been incurred.

Engineers are often surprised to discover that it is important not only to make the right decision as early as possible, but also to make that decision final. The later changes are made, the more expensive they will be. Figure 1-4 shows that the costs for a certain late change to a product grow exponentially with the progress of product development. In a logarithmic scale this appears as a straight line. Here the same change is shown but at different stages of product development.

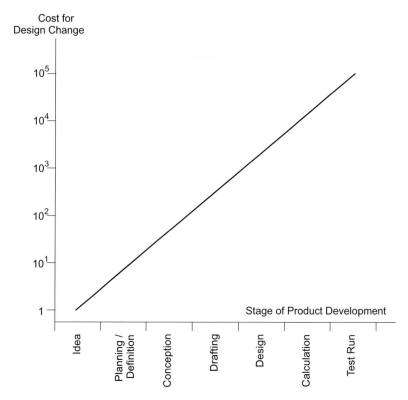

Figure 1-4 The costs for identical changes of a product at various phases of product development

It follows from this not only that changes to the design become more expensive and time-consuming the further the product development has progressed, but also that product deficiencies recognized too late can result in costs that threaten the viability of the entire project. Whoever doubts these findings should bear in mind that the cost for an alteration to a large machine tool can speedily reach several hundreds of thousands of dollars and a recall of automobiles can easily top billions even though the defective part may be valued at mere cents.

To summarize, it follows that the minimizing of product development time is the key management objective, thereby enabling optimal total profit to be achieved and expenditure to be considered a time-dependent variable. In times when people focus on cost reduction, this is an important point.

1.2 Simultaneous Engineering – Concurrent Engineering

1.2.1   Classical Steps of Product Development

The product formation time is the interval between the first conceptualization of a new product and its series production (Figure 1-5). It includes the product development and the production time.

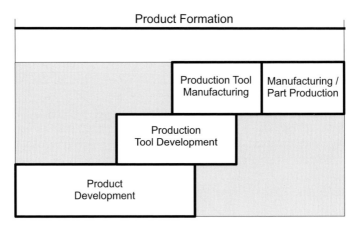

Figure 1-5 Elementary steps of product formation

Idea finding, planning, definition, and draft phases are, depending on the author, part of the product development, or are considered a special, added element. Depending on the definition, product development includes not only the actual development of the product but also the development of the tools necessary for the production, that is, the development and manufacture of the production equipment. Such differences in definition are to be taken into account when comparing the development times for products of different enterprises.

Product development and production equipment development are similar in method. Production equipment development should merely be considered as a special kind of product development, a process for developing the production equipment for the "product." The following paragraphs consider product development in greater depth in terms of a methodical approach. The findings are broadly applicable also to the development of production equipment.

The process of product development is split into different steps by various authors and is not standardized. Peculiarities typical for certain branches are thereby revealed.

Figure 1-6 is based on the VDI Guideline 2221 ff. [VDI 22][1], which defines the product development steps: idea, planning/definition, conception, development/realization.

---

[1]   VDI Verein Deutscher Ingenieure means German Society of Mechanical Engineers

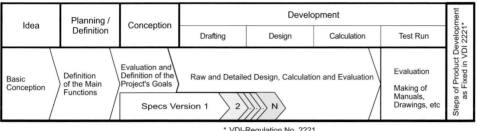

* VDI-Regulation No. 2221
Verein Deutscher Ingenieure
(German Association of Mechanical Engineers)

Figure 1-6 Definition of the classical steps of product development according to VDI

The basic idea and concept will form the essential properties of the finished product. During this phase, and especially during the following phases of planning/definition and conception, the functions, the basic design, and implicitly also the production processes are defined and defined as aims. During this early phase decisions are made that define the resources and processes as well as the costs, making changes impossible or at least very expensive (see Fig. 1-3). The conception phase is followed by the first version of performance specifications. During the following phases of development/realization these guidelines must be optimally implemented. These relationships are illustrated in Fig. 1-6 and are not discussed further at this point.

1.2.2   Requirements for New Product-Development Strategies

The main demands on new strategies is to shorten the time for product development while undercutting the budget and improving the product quality. This involves large development teams from various fields which are allowed to work in separate locations. To fulfil these requirements the most important points influencing the time of development need to be defined, discussed, and formulated in new strategies. The time for product development depends on:

- The fixing of design characteristics early on

- The degree of parallelism in partial processes

- The intensity of information exchange

- The degree of computer integration

- The level of motivation

Fixing of design characteristics early on

For the success of product development it is important to define fixed design characteristics as early as possible. Every design engineer and work team can orientate to clear instructions. If instructions are missing every work team will favor its own design characteristics; such inhomogeneous designs require harmonization later with more effort. The early fixing of design characteristics results in an early fixing of materials and production processes.

Degree of parallelism in partial processes

To shorten the time of product development it is essential to select as high a degree of parallelism as possible. Because parallelism is a discrete process that allows rational parallel ramification only after corresponding partial projects are finished, the organization and monitoring of the total project are highly important. It has to be taken into account that only those partial areas can be parallelized that are completely independent of one another for the duration of the work on the partial project.

Information exchange

A highly effective product development depends on an intensive exchange of information. It is important not only that information is exchanged between work groups, but also how and how often this takes place. Only if information is exchanged speedily and fully is it of any value for the project.

Computer integration

Computer integration is the backbone of an intensive information exchange. Only a high degree of computer integration ensures that all work teams involved are able to refer to the same defined database at any time and that they are able to record their results in the database. Current product developments are so complex that it is impossible to realize these aspects without the help of a computer and a suitable electronic data management (EDM).

Motivation

Teamwork is always effective and successful when motivation is high. Unmotivated employees will always use teamwork to hide within the group and to make others responsible for bad results. Therefore, the question of motivation holds the key to minimizing product development time. There are various approaches for motivating employees but the most important point that decides whether they are motivated or not is how the employees themselves judge the feasibility of the project.

1.2.3   The Principle of Simultaneous Engineering

The demands made on effective product development systems lead to two main points: *concentration and communication*. Expressed in a different way, the demands are:

- A high grade of parallelization of previously sequential developed steps
- A general defined data base available at all times to all team members

Appropriate new strategies for product development have become known as "Simultaneous Engineering" (SE) or "Concurrent Engineering"(CE).

The principle of simultaneous engineering is shown in Fig. 1-7. The upper diagram shows the classical (sequential) development chain in which each step in the development of a product can start only after the preceding one is finished. The basic idea behind simultaneous engineering is that the partial parallelization of these development steps, which were previously carried out in sequence, avoids time-consuming and cost-intensive iterations (indicated in the illustration as marked squares).

While by definition sequential processes work iteratively because they allow the revision of an already finished step only after the next step is finished, simultaneous engineering enables work processes to be carried out in an effectively parallel, or at least partly parallel manner.

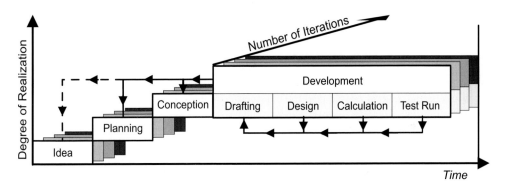

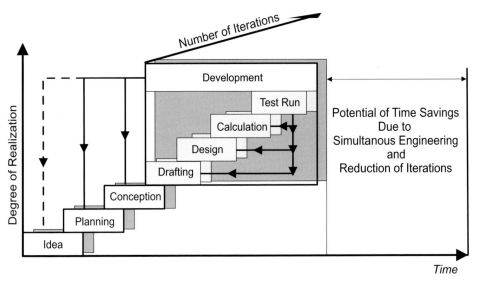

Figure 1-7 Saving potentials in development time by higher degrees of parallelization and the avoidance of iterations

A continual and optimized flow of information between all partners is required for practical parallelization. Therefore, a common, continually updated database available to all team members is an essential prerequisite. The most important element here is the project management, supported by a coordinated control system [electronic data management (EDM)].

New forms of cooperation are being developed to define and organize interdisciplinary partners. Examples are computer mediated communication (CMC) or computer supported cooperative work (CSCW) systems [BUL95].[1]

On this basis parallel work is made possible and it is ensured that the results continually flow back into the further development process. The actually attainable savings depend on the line of business and the product. They vary and, compared with classical models, can save as much as 50% of the costs [ALT94].

The aspect of parallelization is often pushed to center stage in discussions about simultaneous engineering, possibly because it is intellectually understandable and can be presented graphically. Parallelization, however, has its limits. An important point, perhaps the most important, is the aspect of communication and the resulting planning and organizing advantages. An example known to everybody can explain this:

When preparing a meal, peeling the potatoes is a time-determining factor. The preparation takes less time if several people peel the potatoes. Although two people (possibly because of mutual stimulus) manage approximately twice as many as one person, the additional productivity per person decreases with each additional person. As more persons join in the task mutual obstructions increase. The necessary work organization, too, consumes an increasing part of the productivity growth. Employing the chefs as supervisors for the peelers does not improve the productivity either. But the result would be that during the peeling it can be confirmed that the potatoes are suitable for the meal, optimally sized pieces are cut, the necessary amount of water and the time when the cooker should be switched on can already be determined, and so forth. It is the improved communication, not the increased speed, that will make it unnecessary to go over the work again and, for example, cut smaller pieces or even peel other potatoes. Even the doubling of the speed cannot make up for a second round of peeling, especially as, similar to product development, the time of completion and the quality of the meal will be negatively affected.

As the effect of better communication in simultaneous engineering developer teams has such a great influence on the quality of the product and on development time, and because models improve communication tremendously, they are an important contribution to the acceleration and improvement of product development.

In this book the important subject simultaneous engineering is not discussed in further detail beyond the aspect of the necessity of models. For further reference see Ranky [RAN97] and the VDI report 1148 [VDI 11].

---

[1] It seems almost a contradiction that the new computer-supported freedom in product design and product development can be used effectively only if it is controlled by a highly restrictive project management.

1.3 Models

New product strategies take into account that the requirements on products and thereby product development have changed. The following subsections discuss the influence of models and prototypes on the optimal implementation of these new strategies. The observations show that for new product development strategies it is important not only to use models but also to consider how fast they are available, at what step of the development process, and the interactions that occur.

The accepted terms and definitions of classical product development are used.

1.3.1 Model Classes

The demands on models differ according to the degree of progress the product development has reached. It is sensible to agree on a model definition and to assign this to certain steps in the product development irrespective of the question of how these models are produced.

In the relevant literature a large number of various terms and suggestions for model definitions can be found. On the one hand they are often characterized by the planned use and by specialties typical for certain branches, and on the other hand they are often too specifically orientated to rapid prototyping processes.

It would be very useful to apply a very detailed definition as formulated by the German Association of Industrial Designers and Stylists [*Verband der Deutschen Industrie Designer (VDID)*] and the German Counsel for Styling (Rat für Formgebung*)*:

- *Proportional model*

  Shows the outer shape and the most important proportions. Facilitates communication and motivation, supports fast exchange of communication about the intended product properties, enables a fast concensus on the product idea. The production process must be fast, simple and cheap. Disposal and recycling are very important. Proportional models are often called "concept models" or "show-and-tell models."

  Degree of abstraction: high; degree of detailed specification: low; functionalities: none.

- *Ergonomic model*

  Supports the fast decision about feasibility (is it possible to develop this product and should it be done?). Shows important details for operation and use, and also, if applicable, important partial functions.

  Degree of abstraction: medium; degree of detailed specification: medium; functionalities: some.

- *Styling model*

  Shows the outer appearence as close as possible to the (series) sample. Surface finish needs to have "showroom" quality. Supports the fast decision on construction and manufacturing methods. Enables third parties (customers, sales, press, suppliers) to pass their judgments at an early stage, enables public relations work.

Degree of abstraction: low; degree of detailed specification: partially high; functionalities: some.

- *Functional model*

  Enables the proving of the numerical simulation calculations and the early testing of certain functions (how it could be assembled, easy maintenance, kinematics). Shows some or all important functions, if necessary without showing the outer shape. Forms the basis for inquiries by customers and suppliers. Gives relevant information for tool and mold manufacturing, for the construction and installation of the means of production.

  Degree of abstraction: low; degree of detailed specification: high; functionalities: several.

- *Prototype*

  Resembles the (series) sample closely or, if necessary, exactly. Is produced according to production documents. The only difference from the series product lies in the production process. Enables the testing of a single or several product properties (how it can be assembled, ejectablility, start of special approval processes). Enables the production of tools (rapid tooling). Enables the preparation for market introduction by press campaigns.

  Degree of abstraction: none; degree of detailed specification: high; functionalities: all.

- *Sample*

  Already produced in series, possibly a pilot batch, production batch, preproduction, or principal batch. Enables the entire testing of all product properties. Supports the training of production and maintenance personnel, supports the start of mass production, enables the adjustment of production and assembly sequence. Supports the detailed planning of customers and suppliers.

  Degree of abstraction: none; degree of detailed specification: high; functionalities: all.

In the daily rapid prototyping routine this classification is often too detailed. Mostly a simplified model is used, compressing the six model classes of the VDID to just three classes of models (or prototypes).

- *Solid images*

  Covering: Proportional models, ergonomic models and styling models or "show-and-tell" models. Visualizing proportions and general appearance.

- *Geometrical prototypes*

  Testing of handling, operation and use. Visualizing the exact shape including the desired surface qualities.

- *Functional prototypes or models*

  Covering: Functional models, prototypes and samples

Figure 1-8 shows the simplified definition of model classes in relation to the main product development steps as defined in Fig.1-6.

It is important to realize, that "rapid tooling" and "rapid prototyping" as defined in Section 1.4.2 Figure 1-12 is not a model class but an application of rapid prototyping technology. To be used as a tool or to be sold as a product RP parts have to have at least the quality of functional models.

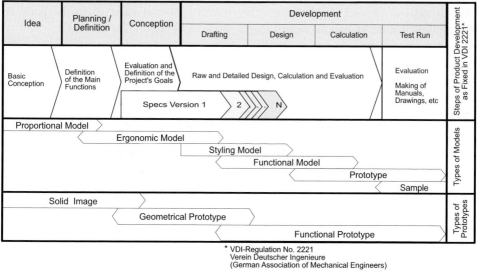

* VDI-Regulation No. 2221
Verein Deutscher Ingenieure
(German Association of Mechanical Engineers)

Figure 1-8 Steps of product development in relation to various model definitions

Although engineers readily agree on the meaning and the terminology of functional models, prototypes, and samples, the classification into proportional models, ergonomical models, and styling models is poorly understood and the value of these models is generally doubted. It is, however, a key process in product development to agree on the product and its general realization.

As this agreement is, even in technically orientated companies, mainly a communicative process, it is especially important to give everyone a fast, clear picture of one's own ideas. If it is not possible to convince all persons involved by means of a drawing or sketch and the following simple model – albeit rough and ready, but one that still gives a general impression – preparation of all further detailed models should be stopped.

The importance of this first step is shown by examples in the field of architecture. Here, the fundamental decision for or against proposals by competing architects is made in principle on the strength of models that, from the point of view of a technician, are rather rough. In this context the notion of a model as a "three-dimensional idea" becomes especially clear.

Only after the proportional model has led to the general decision to proceed with a product idea and the product has been specified by means of ergonomical and design models will the functional models and prototypes become effective and result in a faster and more effective product development. With their help geometrical and kinematic properties especially can be tested speedily and efficiently; they can be correlated with the requirements of the mechanical-technical properties of the product and with the production options.

It is decisive for the success of product development to obtain an overview of all rapid prototyping processes, and the special qualities of each, and to use them powerfully for the relevant steps of product development. Therefore it is not advisable from the point of product development to commit oneself to one single rapid prototyping process.

1.3.2   Influence of Models to Speed Up Product Development

Figure 1-9 shows the principal phase of product development versus time For this discussion a diagram in the form of a first-order lagging function (Fig. 1-9, graph a) is sufficient, although the product is not developed continually but rather with each finished step, and therefore ladder-shaped; furthermore, repeated reworkings of single parts may take the product development some steps back again.

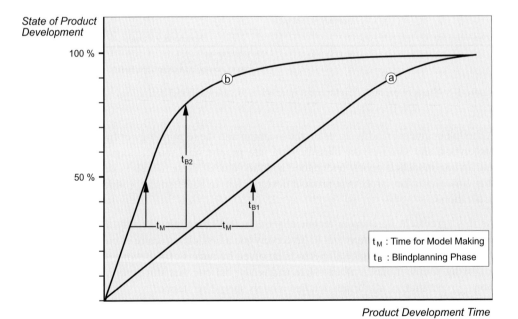

Figure 1-9   Influence of time used for model making (t_M) on the blind planning phase (t_B) as a function of product-development time

The shortening of the product-development time results in a correspondingly steeper graph (Fig. 1-9, graph b). Furthermore, considering the time needed for the production of a model or prototype (t_M), we notice that the product development proceeds "blind" from the time a model maker is comissioned until the model is finished, that is, without verification of the achieved development of the product [blind planning phase (t_B), blind planning risk]. If the model-building time stays unchanged, shortened product-development times result in ever growing sections of product development being affected by a blind planning risk (t_{B1} -> t_{B2}). The basic relationships show clearly that any further shortening of product-development time can be realized only if the production time for models is also shortened, at least proportionally to the product-production time. They also show clearly that even models with constant product-development times will contribute to the improvement of the product because they shorten the blind-planning time.

In addition, Fig. 1-9 makes clear how the use of rapid prototyping or virtual reality (VR) are differentiated. If the methods of virtual reality really do simulate geometrical and later also the physical relationships from various viewpoints (in real time and, if necessary, speeded up) only a few hours after construction, then we have made a great step toward a blind planning time of "zero." This makes VR processes indispensable, especially when viewed in the context of even shorter product-development times. They become the ideal supplement to rapid prototyping model building processes.

- *Cost assessment and cost influencing*

Product development presents us with the dilemma of being decreasingly able to influence costs with growing product development, while at the same time continually improving our cost assessment. The design engineer fixes the costs at the beginning of product development, that is, at a moment when he is not yet able to assess the costs reliably. His ability to influence the costs decreases as he is better able to assess the costs in the course of progressing realization of the product (Fig. 1-10).

It is essential therefore to find methods that tend to push the graph "cost assessment" above the graph "cost influencing."

This necessity cannot be translated into practice completely, but by using models the tendency is in the right direction and the intersection of the two graphs drifts to the left, that is, to an earlier point in time in product development. Figure 1-10 shows, integrated in a classical diagram according to Beitz et al. [BEI94], the influence models have on the cost assessment.

By using models costs can be assessed earlier and at a time when the resulting total expenditures can still be effectively influenced

- *Influence of models on critical success factors*

The requirements discussed in Section 1.1.2 can be better fulfilled if the product formation is supported by models and prototypes. The following positive influences result when combining the model classes:

Binding interactions between marketing, developing and management are fostered by "solid images", also called "show-and-tell" models, even if the outlines are sometimes rather rough and only the general features are recognizable. This applies even more as it must often be assumed that proposals forwarded by development engineers in the form of technical drawings are not, or not completely, understood by marketing and management people. On the other hand, a physical model, simplified in line with the product concerned and – if possible – on the original scale, can be safely judged by anyone as regards design and styling.

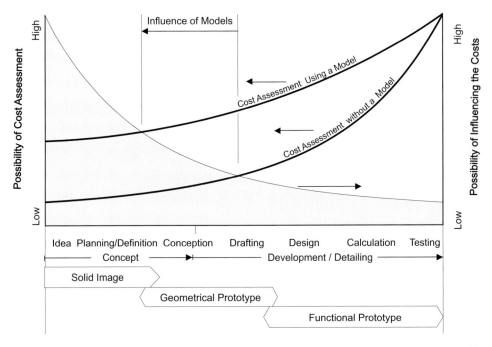

Figure 1-10 Cost assessment and cost influencing as functions of product development progress with and without the use of models

Decreasing product life spans, increasing downward pressure on prices, and the demand for interdisciplinary cooperation conglomerate with the problem of shortened product developing time and fast, smooth communication. Models and prototypes improve the quality of communication. Experts of various disciplines can study each stage of development from their own specific points of view and are able to make changes at an early stage. For example, orifices, insignificant for the technical designer, could mean that the product cannot be cast. This group of models is known as functional prototypes because with their help, singular functions can be tested.

The various standards and regulations that need to be met, and that may differ locally, raise the questions of certification, inspection and approval. If, in the course of product development, for certain test purposes specialized prototypes are placed at the disposal of the approving authorities as test samples, these procedures may be accelerated and optimized (technical prototypes or samples).

The requirements of environmental protection are expressed in demands for recycling and disposal. For example, disassembling or environmentally compatible packaging tests can be carried out at an early stage with the assistance of models. For these purposes also technical prototypes and samples are needed.

Models are consequently important aids that influence positively the communication within development teams and between various teams, and also the overall motivation, the early recognition of faults, and the early review of concepts.

1.4 Model Making by Rapid Prototyping as an Element of Simultaneous Engineering

The previous discussion demonstrates the value of models for faster and more effective product generation. In these observations it was not important to look at the way the models were created. The following subsection considers the characteristics of rapid prototyping models and asks why models made by this method are especially suited to support effectively the strategies of fast product development and product generation.

1.4.1   Rapid Prototyping Models as Guarantors of a Defined Database

Even when taking the relationships discussed in Section 1.2.3 into account an important link is still missing for the implementation of simultaneous engineering into the classical chain of development: The continuous, universally available and defined database does not exist as long as the models are made manually or semimanually.

The conventional model making process, however, does rely on the joint database in the form of two-dimensional (2D) drawings and sketches, but it usually changes the geometry in the course of necessary "interpretations" of the 2D drawings during the model making. This step is often taken deliberately to create the final geometrical form which is then measured and returned to the design process. This means that during the time when the model is being made there exists no defined database to which other members of the team could refer. Simultaneous engineering is therefore not feasible for the duration of the model making.

In the rapid prototyping process the complete 3D geometric information from the CAD is split into layer information and the layers are built directly with the aid of the computer. Rapid prototyping processes build 3D models, therefore, only on the basis of 3D CAD computer data. Using the data they receive, they merely reproduce the CAD design three-dimensionally and without changing it physically. The database, compulsory for everyone, is thereby retained during the model-making process. In this way the rapid prototyping processes form the final link in the chain of computer integrated manufacturing (CIM) and they provide the basis for the realization of CIM strategies and their integrated methods of simultaneous engineering. Figure 1-11 gives a clear picture of the interrelationships:

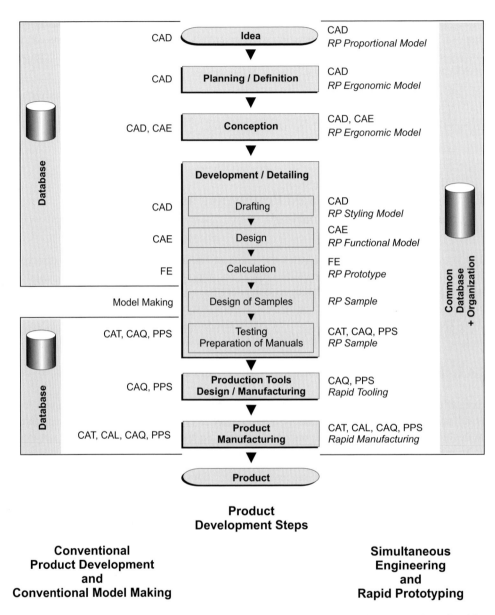

Figure 1-11   Partly closed CIM-chain *(left)* and completely closed chain by implementation of rapid prototyping processes *(right)*

The single steps of product development, from the idea to the finished product, are arranged sequentially for better understanding. The definition of terms and the interaction are arbitrary and serve only as explanation. A line for conventional product development (Fig. 1-11, left) and another line for simultaneous engineering (Fig. 1-11, right) and the pertinent CAx components are shown. It can be clearly seen how the conventional model making interrupts the data flow, denying a closed CIM structure. If rapid prototyping techniques are employed in simultaneous engineering, however, closed data structures become visible that are vertically compatible and that allow all members of the product development team to have access to the current data record at all times. It also becomes clear that, in contrast to classical model making, rapid prototyping processes are not restricted to one (or more) model making phase(s) but that they can be used randomly for various model characteristics at any point in the product generating process.

1.4.2 Definition: Rapid Prototyping – Rapid Tooling – Rapid Manufacturing

All processes by which 3D models and components are produced additively, that is, by fitting or mounting volume elements together (voxels or layers) are called *generative production processes*. This precise generic term is seldom used in practice. More common are various terms that relate to the single aspects of the component. Solid freeform manufacturing (SFM), sometimes called solid freeform fabrication (SFF), emphasizes the ability to produce framed solids by means of free form surfaces; desktop manufacturing (DMF) enables components to be produced in an office environment (on the table). A number of these terms can be found in specialized literature on the subject and mostly appear in the form of three-letter abbreviations and are more often confusing than explanatory (those used most often are explained in the text).

Each term in use has its justification, from the point of view of its author at least, and is a suitable expression for the specific purpose it is used for and cannot be exchanged against any other term. However, in this book the term *rapid prototyping* is used deliberately and constantly as a generic term. The expression "rapid prototyping process" is certainly not the best, maybe even one of the worst terms available to us. Rapid prototyping tells us nothing when closely analyzed. "Rapid" is relative. It gains quality only when it tells us "faster than what" or at least "how fast." There is also a certain danger in using the term "rapid": it could mean that these processes are intrinsically faster than others. This is not necessarily so. There is no general rule to be found here. The speed of rapid prototyping processes depends to a great extent on the geometry. Whoever needs only a board of $25 \cdot 25 \cdot 1$ inch is better served by the semifinished product and a saw. No computer-aided model making process will be faster.

The word "prototyping" is also inapt because many applications of computer-aided production processes do not deal with the production of prototypes in the strict sense. Apart from design models and demonstration models, molds and tools are made and even (small) series are produced.

The term *rapid prototyping* has, however, an unbeatable practical advantage. It is engraved in everyone's memory. It is viewed as a synonym for computer-controlled and therefore automatic generative processes. Rapid prototyping together with its most prominent member, stereolithography, are well known in this combination. They are self-explanatory and thereby fulfill the most important requirements of a standard term.

In contrast, most of the other terms used and explained in the text or in the appendix require additional explanation by the user: "That is something like rapid prototyping." For this the reason we call this process "rapid prototyping" right from the start.

The terms rapid tooling and rapid manufacturing are subordinate to that of rapid prototyping and relate to special uses and areas of application.

Rapid prototyping encompasses the science of generative production processes and is therefore a technology. Owing to their newness, some applications acquire their own terminology for techniques or strategies (often called "applications" as well) which are often used synonymously with those of the technology. Whereas in milling machine technology there is no differentiation in method between producing positives and negatives, this is definitely the case in rapid prototyping technology where compulsory standards are only partially available. It is, however, very important to realize that the application of rapid prototyping technology is, methodically, really a technique. Therefore, concept models and geometric prototypes (solid imaging) as well as functional prototypes and technical prototypes (functional prototyping) on the one hand, and generative tool making (rapid tooling) and generative series production (rapid manufacturing) on the other hand, adopt the status of a strategy irrespective of their practical significance.

Depending on the architecture of the machine and the material used, the application of rapid prototyping technology leads to solid images or concept models/geometry prototypes or to functional prototypes/technical prototypes. Applications especially for metal materials have brought about the development of generative tool making (rapid tooling) and generative series production (rapid manufacturing).

Prototypers[1] used for the production of solid images or concept models/geometry prototypes (concept modeler) and those used for the production of functional prototypes/technical prototypes are technologically similar. Whereas concept modelers are suitable for the production of relatively rough but cheap models, functional prototypers produce more complex, more detailed, and more precise – but also more expensive – models.

Because rapid prototyping processes are practically unlimited in their ability to form complex shapes, they can produce both positives and negatives. Negatives are produced as dies or molds (die or mold inserts, respectively) for preproduction or small-batch production with corresponding positives. In this case, it is called generative tool making or *rapid tooling*. Rapid tooling is, therefore, of special importance because the "step into the tool" is very time consuming, prone to faults, and expensive for all product generating processes.

The terms generative series production or *rapid manufacturing* (also: rapid production) assume that rapid prototyping methods can be used directly for the production of all kinds of (mass) products. This is already being done with special applications such as, for example, medical implants (CP-GmbH, Germany) or plastic aligners for straightening adult teeth (Align Technology, CA, USA).

---

[1] Prototypers are machines or installations that produce prototypes. The term was coined during a discussion at a technical meeting with Professor E. Schmachtenberg, IKV-RWTH Aachen, Germany, at the Euromold fair at Frankfort, Germany, 1995.

Figure 1-12 summarizes the various applications of rapid prototyping technology with respect to specialization and the similarity to the later mass product of the models thereby produced.

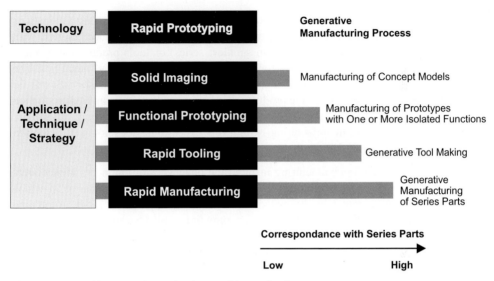

Figure 1-12 Rapid prototyping technology and its applications

If we follow the actual terminology the following definitions result:

- *Rapid prototyping*

  Rapid prototyping describes the technology of generative production processes.

Solid imaging and functional prototyping describe the applications of rapid prototyping technology. Solid imaging includes the production of relatively simple, mechanical-technological nonresilient models that nonetheless display the outer form and the features of the final component relatively well.

Functional prototyping is the application of rapid prototyping technology to prototypes made of plastic, metal, or other materials that simulate one or more mechanical-technological functionalities of the final series component.

In many cases solid imaging and functional prototyping often become the time-determining factor during the first phase of product development.

- *Rapid tooling*

  Rapid tooling describes those applications that are aimed at making tools and molds for the production of prototypes and preseries products by using the same processes as those used in rapid prototyping. This concerns both the model (positive) as well as the mold (negative). Anglophones talk here of "pattern making" and of "mold making."

Against this background, rapid tooling becomes the time-determining factor in the second phase of product development, that of optimizing the actual product, developing the means of production, and the production itself.

- *Rapid manufacturing*

  By rapid manufacturing or rapid production we understand rapid prototyping applications that produce products with serial character. These can be positives produced directly with rapid prototyping methods (e.g., plugs in smallest series) or tools produced with rapid prototyping processes usable directly for the production of the required quantities. The mechanical-technological properties of today's rapid materials are in most cases still far from the target characteristics of the products. For larger production quantities production times are still relatively lengthy. For these reasons rapid prototyping is usually uneconomical and rapid manufacturing, with a few exceptions, does not (yet) belong to the production processes that are in use.

The possibilities of rapid manufacturing inspired the phantasies of engineers immediately after details of the first rapid prototyping processes were published. Scenarios in which spare-part stocks are completely eliminated and replaced by appropriate rapid prototyping installations "just in time" have been known for some time now. One suggestion of making the entire stock-keeping of naval units, for example on an aircraft carrier, superfluous while simultaneously guaranteeing a flexible provision by using appropriate rapid prototyping (metal) installations was especially discussed in detail.

Other scenarios in which the use of rapid manufacturing methods alone can enable the transport of tools and spare parts to distant celestial bodies such as Mars are also being seriously discussed at present. These reflections are of value only, however, if the prototypers are able to work with the materials available there.

Even if these scenarios still seem unrealistic today, recognizable development trends make such applications ever more probable.

Rapid manufacturing as a tool of "customized mass production" processes will gain more importance in future in view of the following development trends in rapid prototyping processes and the demands made on the products:

- shorter product life time,
- increasing product complexity,
- growing individuality of products,
- smaller series.

1.4.3   Relating Rapid Prototyping Models to Product Development Phases

Figure 1-13 shows the relationships of rapid prototyping, rapid tooling and rapid manufacturing to the basic product development phases.

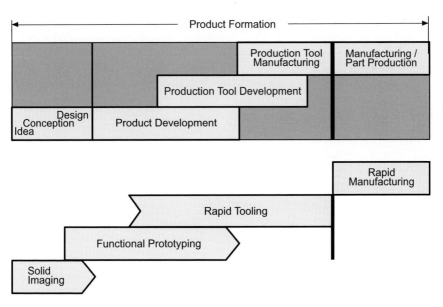

Figure 1-13   Relating of rapid prototyping, rapid tooling and rapid manufacturing processes to the basic product development phases

2 Characteristics of Generative Manufacturing Processes

Industrial rapid prototyping systems on the market today are subject to a high development speed. New processes still in the laboratory stage or under development today will break into the market. At the same time, well tried and tested systems will be upgraded within a relatively short time.

As the equipment presently on the market will be obsolete or approaching obsolence over relatively short periods the physical-technological bases of the various processes are portrayed and discussed in detail in this chapter. Chapter 3 then shows which industrially offered installations derive from which fundamental processes. This procedure not only facilitates the assessment of the current processes, but it also supplies the basis for the assessment of future industrial processes.

This division is also intended to separate the representation of basic principles valid for a longer term from machine concepts that change more quickly. In reality, however, overlaps and repetitions are unavoidable.

2.1 Basic Principles of Rapid Prototyping Processes

Rapid prototyping processes belong to the generative (or additive) production processes. In contrast to abrasive (or subtractive) processes such as lathing, milling, drilling, grinding, eroding, and so forth in which the form is shaped by removing material, in rapid prototyping the component is formed by joining volume elements.

All industrially relevant rapid prototyping processes work in layers. Like the half-breadth-plan of a ship, known from classical model making, single layers are produced and joined to a component.

In the strict sense, rapid prototyping processes are therefore 2½D processes, that is stacked up 2D contours with constant thickness. The layer is shaped (contoured) in an (x-y) plane two-dimensionally. The third dimension results from single layers being stacked up on top of each other, but not as a continuous z-coordinate. The models are therefore three-dimensional parts, very exact on the build plane (x-y direction) and owing to the described procedure the stepped in the z-direction whereby the smaller the z-stepping is, the more the model looks like the original.

Although all rapid prototyping processes known today work in this way as 2½D processes, some processes (e.g., extrusion processes) are in principle 3D processes, which means they can add incremental volume elements at any chosen point of the model.

The special characteristic feature of rapid prototyping processes is that the physical models are produced directly from computer data. In principle it is thereby unimportant whence the data are provided as long as they describe a 3D volume completely. Data from CAD design, from the processing of measurings and reverse engineering or other measurements [computer tomography (CT), magnetic resonance tomography (MRT)] may be used equally well.

In this way model making has become an integral part of the computer-integrated product development. From the product development aspect rapid prototyping models can, therefore, be regarded as three-dimensional plots or facsimiles of the corresponding CAD data. The decisive advantage in contrast to classical manual or semiautomatical model-making processes lies in the fact that the data remain unaltered by the model making. As a result no data need to be taken from the model. Because the making of rapid prototyping models does not alter the common database, rapid prototyping processes have become the most important elements of modern product development strategies such as simultaneous engineering.

"Rapid prototyping" or "generative manufacturing processes" are classified into two fundamental process steps:

- Generation of the mathematical layer information (Section 2.2)

- Generation of the physical layer model (Section 2.3)

The generation of layer information is based on a purely computer-orientated CAD model (Figure 2-1). The CAD model is cut into layers by mathematical methods. This layer information is used for the generation of physical single layers in a rapid prototyping installation, the so-called prototyper. The total sum of the single layers forms the physical model.

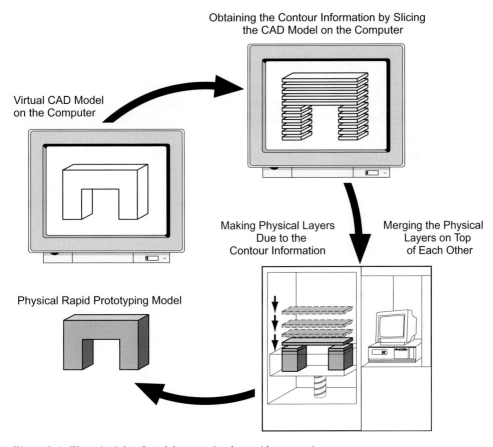

Obtaining the Contour Information by Slicing
the CAD Model on the Computer

Virtual CAD Model
on the Computer

Making Physical Layers
Due to the
Contour Information

Merging the Physical
Layers on Top
of Each Other

Physical Rapid Prototyping Model

Figure 2-1   The principle of model generation by rapid prototyping

2.2 Generation of Layer Information

The process whereby layer information is generated, thereby providing the geometrical raw
data for the production of the physical model, is divided into three steps:

- Description of the geometry by a 3D data record (Section 2.2.1)
- Generation of the geometrical information of each layer (Section 2.2.2)
- Projection of the geometrical layer information on each layer (Section 2.2.3).

2.2.1 Description of the Geometry by a 3D Data Record

2.2.1.1 Data Flow

The production of models and prototypes by means of rapid prototyping processes requires that the geometry of the component is available as a 3D data record. This is achieved in most industrial applications by construction on a 3D CAD system. The data are produced independently of the prototyper and need to be prepared and transferred via interface. To build the prototypes further process- and installation-specific calculations are necessary in addition to geometrical data. An examination of the data flow reveals that there are two areas: one concerning the design, which is covered by the general CAD software and another one concerning the rapid prototyping installation, which requires special Rapid Prototyping software Figure (2-2).

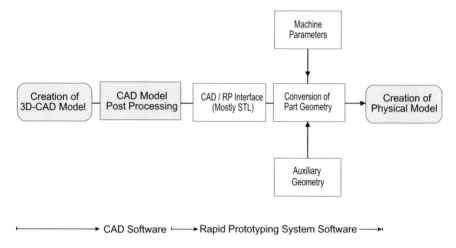

Figure 2-2 Data flow in rapid prototyping

As shown in Figure 2-2 the geometrical data are transferred from the CAD model via a neutral interface (e.g., STL, Section 2.2.2) to the installation. To proceed with the production of a rapid prototyping model it is first necessary to establish the auxiliary geometries (supports, walls, etc.) required for the construction and to establish the machine parameters required for the control of the rapid prototyping process. This is achieved with the aid of a special rapid prototyping software – the so-called system software.

The basis for all rapid prototyping processes is, therefore, a complete 3D CAD (volume) model.

In practice, attempts are often made at building perfect models from inadequate data; failure is blamed on the data conversion or even on the rapid prototyping. It is obvious that rapid prototyping models can never be more exact than the data on which they are based. This leads to the fundamental condition for using rapid prototyping processes:

Whoever wishes to use rapid prototyping processes for product development must always employ a 3D (volume) CAD.

In addition there are several other possibilities for obtaining data for model making. Figure 2-3 shows the possible data flow for rapid prototyping. It is independent of the task to be achieved and of the process chosen for the model making.

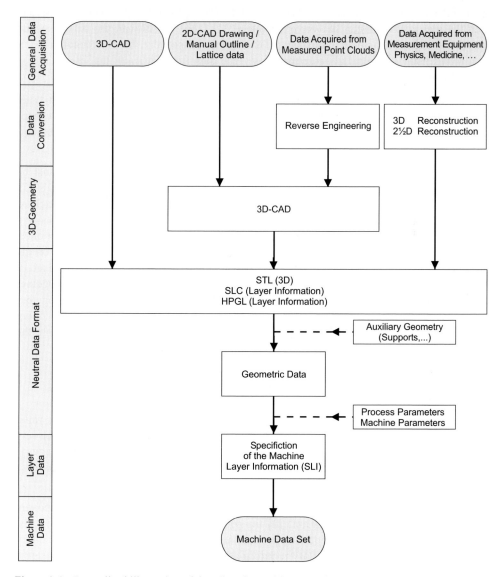

Figure 2-3   Generalized illustration of data flow for rapid prototyping

In addition to 3D CAD designs, 2D CAD sketches, manual sketches, and anything similar can in general be used as input data. To transform them to rapid prototyping models they will require a final conversion into 3D volume information via various intermediate operations. If a body is described only by means of 2D elements then they require conversion into a 3D format by preparing them with a 3D CAD. In most cases this is done using a manual 3D CAD drawing which is no less time consuming than a new design.

If, on the other hand, 3D measurement data exist, for example, from a 3D measuring device, then the measured data can be converted into a 3D CAD model with the aid of special program systems. In the field of mechanical engineering, especially, this conversion of point data (considered geometrically inferior) to surface data (geometrically higher quality) is called "reverse engineering."

The conversion of measured data in the form of point clouds directly into solid body descriptions in a neutral data format, possible in principle, should be done only in exceptional cases. It is not usually possible to relate point clouds clearly even to simple geometrical bodies, and they contain an enormous amount of data.

In any case, the geometrical information of the entire body or of single layers must be converted for transfer via interface into a neutral format [stereolithography language (STL), SLC which is a entire 3D-Systems or Stratasys slice contour format, Hewlett Packard graphic language (HPGL), etc.] as only then is the access to different rapid prototyping processes secured. The generation of auxiliary geometries such as supports and similar ones, which are not necessary with every rapid prototyping process, is done – depending on the process – either together with the generation of geometrical data or separately with the aid of rapid prototyping software. Finally, all data, the geometry and the supports, are together sliced into layers by mathematical means and provide the layer information that, together with machine specific parameters, is necessary for the production of each layer. Depending on the process, the layer information is either completely calculated and stored before the process is started, or it is calculated for each layer simultaneously with the build.

2.2.1.2 Modeling 3D Bodies in a Computer by Means of 3D CAD

The creation of a 3D body is the indispensable prerequisite for the production of a rapid prototyping model. Therefore, the application of prototyping is linked especially closely with CAD processes. For this reason 3D CAD processes will be looked into only as far as is absolutely necessary for the understanding of the fundamental relationships in the production of rapid prototyping models.

Every CAD system uses certain data elements and data structures to describe a component in detail. The data record includes not only the component geometry but also the materials, the quality of the surface, the production process, and much more. The component geometry therefore comprises only one part of the information. The complete information registered in the database of a CAD system for a component is called a CAD model (the product to be made). If the geometric description of a component is 3D then it is called a 3D CAD model.

By choosing a certain CAD system the user commits himself to its database. The structure and the data elements decide to a high degree the quality of a CAD system and its compatibility with other systems via interface.

CAD Model Types

CAD models are defined by model types regardless of the kind of CAD system (Figure 2-4).

The *corner model* defined by points is of less practical importance. It is used, for example as a intermediate model for the semiautomatic transformation of grid data or of 2D CAD models into 3D CAD models.

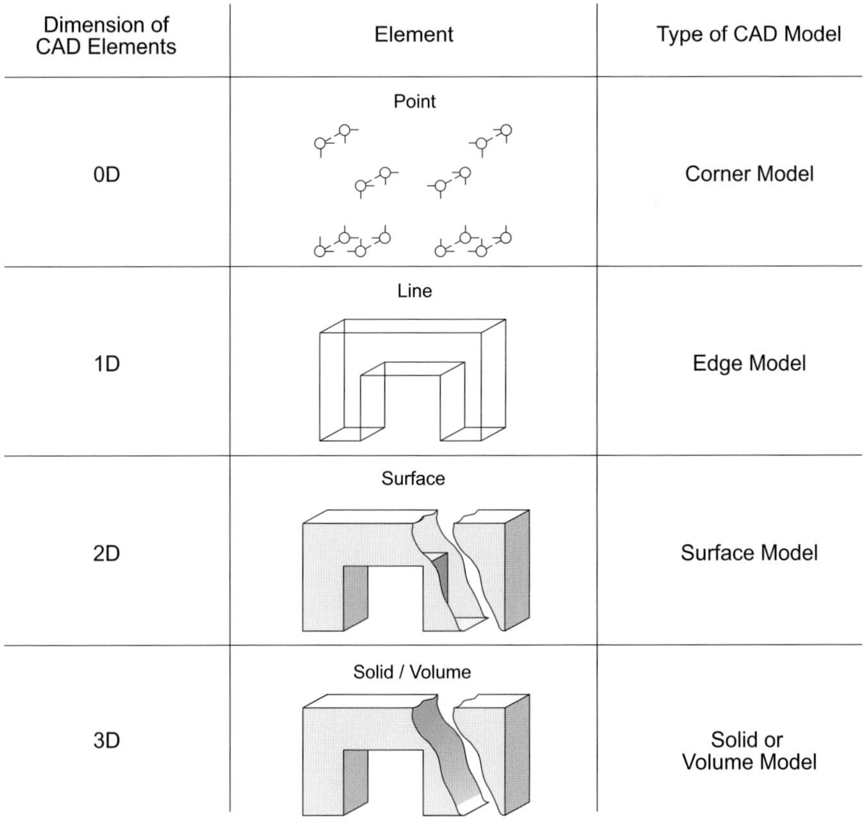

Dimension of CAD Elements	Element	Type of CAD Model
0D	Point	Corner Model
1D	Line	Edge Model
2D	Surface	Surface Model
3D	Solid / Volume	Solid or Volume Model

Figure 2-4 CAD elements and model types

The *edge model* too is more of historical interest today in regard to rapid prototyping. Owing to its small amount of data it enables a fast graphic representation of 3D elements even with low computer performance. Its importance is therefore growing again in connection with virtual reality (VR) applications and digital mockup (DMU). The most important disadvantage of the edge model is the missing information about the exact position of the surfaces and the volumes. For this reason it cannot be recommended as a basis for the production of rapid prototyping models.

All CAD-systems that process components as *surface models* in their geometrical databases are in principle suitable for the issuing of data via a rapid prototyping interface. When a component is defined by its external surface, the user is usually able to calculate the exact component volume as well. This is usually achieved by appointing and storing an additional normal vector for each surface pointing away from the inside of the component. For the complete description of a component therefore it is absolutely necessary that the orientation of the component volume is known.

Solids are optimal for the modeling of CAD models that (among other things) are also used for rapid prototyping. The orientation of the volume is preset exactly and need not be explicitly appointed by the user. Solids are differentiated into:

- Basic solids
- Surface determinating models
- Hybrid models.

For basic solids the component is reproduced in the CAD system by combining basic bodies (so-called geometric primitives) such as cuboids, globes, and cylinders by means of Booleian operation. In most cases more complicated basic bodies can also be used such as those resulting from arbitrary contours revolving around an axis. Figure 2-5 shows both the geometry as well as the constructive solid geometry (CSG) tree of a component that was formed by the addition of two basic bodies followed by the subtraction of a third basic body. The CSG tree is part of the database and reflects the component's history of origin.

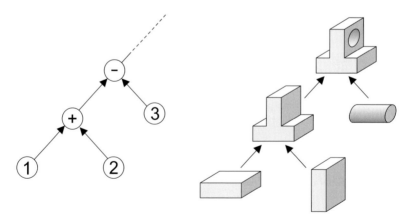

Figure 2-5 CSG tree of a basic solid

The basic solid includes the CSG-tree but it does not contain any information about the single surfaces. If a basic solid is issued as an STL file the contours of the basic body are calculated as a first step. Mistakes caused by inexactitudes in fitting neighboring surfaces are therefore impossible.

With surface determinating models in the extreme case only the details concerning single surfaces and the position of the volume are stored. The position of the volume, also defined by a normal vector standing on each surface and pointing to the outside, does not need to be defined by the user. This model type also enables bodies with extremely complex outlines to be described, which is otherwise possible only with the aid of surface models.

CAD systems do not usually work on the sole basis of one of the two described solids but rather in a combination of both model types. The advantages of both model types are combined in so-called hybrid models, which include elements of the basic solid as well as those of the surface model. CAD systems that work with hybrid models usually generate faultless STL data, as the edges of the surfaces used for the issuing are exactly fitted by the system itself when the resulting hybrid model is generated.

Design Systematics of 3D CAD Systems

Those working for the first time with 3D CAD programs in connection with rapid proto-typing can see from Figure 2-6 how the systematics of modeling with the aid of a parametric volume modeling 3D CAD process (hybrid process) appear in its most important steps. The different basic strategy to that of 2D-CAD can be clearly seen.

Specific 3D CAD systems and their uses are not discussed here.

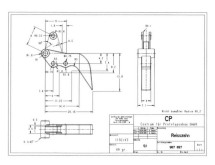

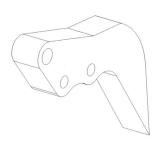

1. Basic Geometry as from a 2D-CAD Drawing   2. Basic Volume after Deepth Allocation

Figure 2-6 (1, 2)

3. Addition of Second Volume (Profiled Solid)   4. Addition of Third Volume

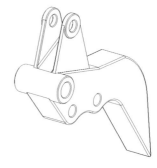

5. Addition of Fourth Volume ....and so on   ... Addition of Details (Chamfers, Radii, Curvatures)

Figure 2-6 (3–6) Selected steps for the 3D volume modeling (Pro Engineer) of an excavator bucket
        tooth

Common Mistakes – Interrupted Contours and Incorrect Orientation of Surfaces

A common mistake in STL data records derived from surface models is the incorrect orientation of surfaces. A normal vector pointing to the inside of the component could lead to problems while generating machine data records, as the interior and the exterior of the component to be produced may not always be differentiated. Surfaces that are not edged exactly inevitably result in gaps in the STL data record. These faults are insignificant in visualizing components during product development and for milling work with tool diameters of millimeter sizes. When using laser beams with diameters of several tenths of a millimeter, all gaps need to be closed at the very latest when the machine-specific layer information is generated (Figure 2-7). This is known as "repairing" the data record and it is done either with the aid of rapid prototyping software (Section 3.1) or manually.

Figure 2-7a shows an example in which repair is still easily achieved, but Figure 2-7b shows a case in which the layer data will almost certainly not be retrievable and faulty components will consequently be produced.

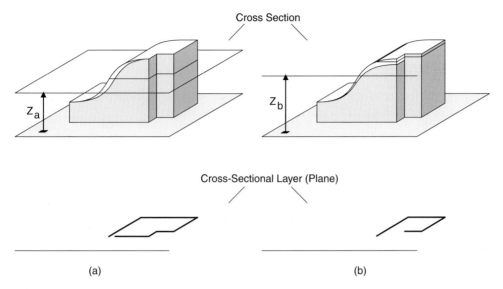

Figure 2-7 Influence of interrupted contours on the generation of layer models

Many CAD systems working with surface models include test programs that are able to test the quality of the rapid prototyping data before or during the production.

Also, CAD systems exist that generate no STL data record at all if there are gaps in the data record or other faults. As the user – especially if he possesses no deeper knowledge of CAD systems – is unable to proceed with his work, it is in practice often better to use a CAD system that in such cases generates a faulty data record rather than no data record at all.

When dealing with difficult or faulty data, use of surface models often offers an advantage. They usually allow single surfaces to be also processed as rapid prototyping data records. This permits the required model to be assembled from its partial surfaces and then worked over in detail.

All the afore mentioned possibilities of repair and management of inadequate raw data portray special cases that should normally be avoided. The easiest way of achieving this is by explaining the entire compulsory procedure of data preparation and data transfer by means of a simple sample before the model making is commenced. On the basis of these preliminary investigations faults in the routine work that follows can be reliably avoided.

Demands on CAD Systems

When judging CAD systems as tools for the realization of the principle of simultaneous engineering and as the basis for rapid prototyping processes some requirements should be taken into consideration:

- *Parametric 3D designs*

  Instead of using fixed measurements, parameters are agreed on that in addition can be correlated with each other by mathematical functions.

- *Hybrid models*

  Hybrid models combine the advantages of both, solids and surface models and are therefore very well suited for Rapid Prototyping applications.

- *Continuous database*

  The CAD system and all associated modules must always refer to a common, compulsory database.

- *Redundancy avoidance*

  A continuous database will avoid redundancies, that is, the database will be free of data unnecessarily stored several times. Such multiple storages are to be avoided in view of storage capacity, speed, and clarity of the program.

- *Open system*

  It needs to be guaranteed that the systems can be linked with specialized modules from independent producers (reverse engineering, CT modeling, rapid prototyping software) (see also interfaces, Section 3.1).

- *Associativity*

  The internal architecture of the CAD systems must ensure that any alterations cause all dependencies to be checked and modified where necessary.

The most suitable CAD system for rapid prototyping should possess as many of the properties discussed in the preceding as possible. The basic requirement must be that it is a 3D volume modeler and that it has an STL or an SLC interface (see Figure 3-1). The transformation via neutral interfaces into another CAD-system and from there via an STL interface to the rapid prototyping machine should always be regarded as a detour. Further, when choosing a CAD system it should, for example, be assessed how easy it is to learn to work with it, and its level of support in performing certain tasks, which may be specific for special branches.

The market offers a vast number of CAD systems that often differ only in specialties typical for specific fields of activity. It is increasingly probable that all important CAD systems either possess an STL interface already or will obtain one in the near future. Usually other complementary rapid prototyping interfaces are available, for example, SLC or HPGL.

Appendix A2-1 includes a list of CAD systems that possess a rapid prototyping interface. The special properties of the CAD systems listed are deliberately omitted. The Internet addresses provided will ensure that the interested reader can inform himself of the current situation speedily and reliably.

Reverse Engineering

Often it is advantageous during construction to stack onto already existing products either entirely or partly. For this reason the entire surface needs to be measured and recorded and returned to the CAD system (reverse engineering). The measurement produces an enormous number of data that are also known as point clouds. Reverse engineering enables these point clouds to be defined as surface elements, thereby facilitating their further processing in the CAD system. However, as a rule, the CAD systems are unable to process the myriad of data. In reverse engineering, therefore, independent modules are used that either interact with existing CAD systems or work as independent program systems. In the second case, only the results are transferred to the CAD-system via a suitable interface.

The subject of reverse engineering is far too complex to be discussed exhaustively here. It will be mentioned briefly owing to its special significance for prototyping in specific areas. It should be taken into account too that the results of reverse engineering often depend on the type of task demanded; certain program systems may, therefore, suit different tasks.

The measurement of an existing geometry is advantageous if one or more of the following work steps dominate the design:

- Making of mirror-symmetrical CAD models from hand models that are never exactly symmetrical

- Making of mirror symmetrical CAD models from semimodels

- Introduction of additional geometries such as grooves or drillings in complex components

- Alterations of the model to enable variations to be probed on the screen, for example

- Isolation of one element of a group of components as a basis for further construction work

These manipulations can usually be implemented easier and faster with the aid of a 3D CAD model and a rapid prototyping model than with a real component. In extreme cases these alterations would not be economically possible on a real component using conventional means. This is all the more true the more subsequent series production processes differ from conventional prototype constructions with their cutting and noncutting shaping processes.

Measured Data Logging

Simple geometries can be recorded manually by means of meter rules and gauges. Complexer geometries and the growing requirements on the accuracy of measurements explain the introduction of automatic and semiautomatic measurement systems. Figure 2-8 shows how different measurement data can be introduced into a 3D CAD model. Important, but not quantifiable, is the subjective influence of the user which will also be reflected in the quality of the 3D CAD model.

The classification of models and the decision concerning the measuring methods used for the measurement of a real model depend, among other things, on the complexity of the geometries to be measured. A differentiation is made between the complexity of single geometries such as, for example, those of single surfaces or smaller surface connections, and the complexity of total geometries, that is, whole bodies. In Figure 2-9 degrees of complexity for bodies are defined.

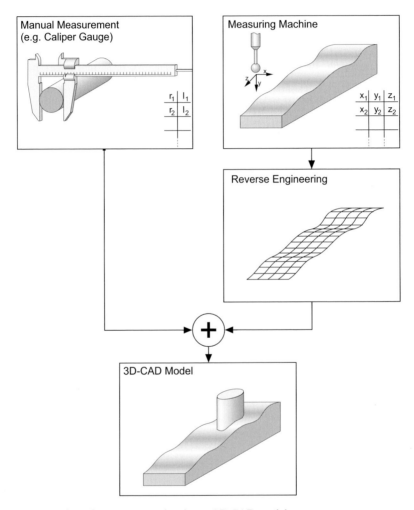

Figure 2-8 Conversion of measurement data into a 3D CAD model

Parallel to establishing the degree of complexity of the body to be measured, the following criteria help to decide whether a metrological recording followed by reversed engineering, in contrast to a direct (new) construction, is preferable or not:

Grade of Complexity	Example	Notice
1	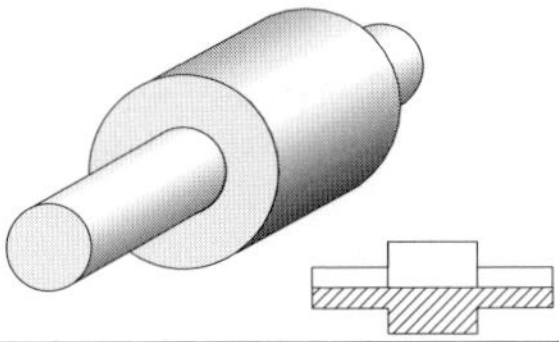	a) Measurment of Simple Surfaces b) Allows only One Plane or Quasi-Plane Reference Face c) Covers Only One Measuring Direction
2	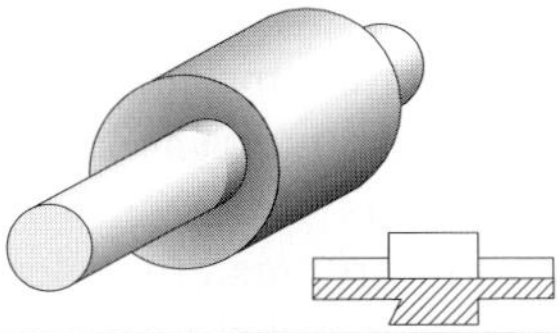	a) Measurement of Complex Surfaces b) Allows Divers Plane or Quasi-Plane Reference Faces c) Covers Diverse Measuring Directions
3	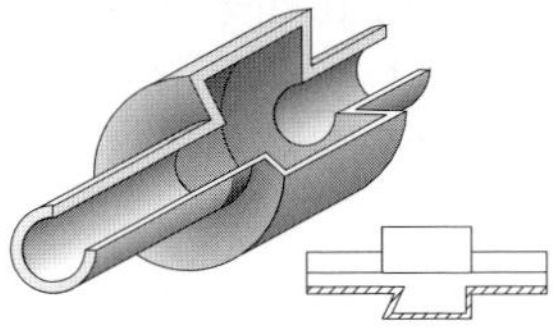	a) Measurement of Volumes b) Allows Divers Plane or Quasi-Plane Reference Faces c) Covers Diverse Measuring Directions

Figure 2-9 Definition of the degree of complexity for 3D bodies

Complexity of Single geometry

- Are these surfaces, at least partly, geometric basic bodies such as, for example, cuboids or cylinders?

- How large is their proportion?

- Do the surfaces join one another only evenly or also unevenly?

- Do the surfaces meet at obtuse-angled or acute-angled joints?

If the degree of complexity of all single geometries is extremely low then the resulting complete model will also possess a low degree of complexity. Such bodies could also be measured completely in the conventional manner.

Measurement Complexity

- Necessary accuracy of measurements

- Size and weight of the object to be measured

- Material and surface condition of the object to be measured

- Costs of measurement reading and processing

- Time-optimized semimanual reverse engineering

Usually it is only a combination of various measurement techniques that leads to a 3D CAD model within justifiably temporal and economical limits. As shown in Fig. 2.8, measurement data may be quite easily connected in a CAD system to a 3D CAD model under very low expenditure if complex areas only are converted to surfaces by reverse engineering, whereas the geometries of simple areas are produced directly and interactively after manual measurement.

2.2.2 Generation of Geometrical Layer Information on Single Layers

To produce three-dimensional models by layer orientated rapid prototyping processes, the 3D CAD solid must be mathematically split into the same layers as those produced physically by the prototyper. This process is known as "slicing." There are two basic methods of doing this. Either the surface is covered with very small triangles, thereby enabling real 3D geometry to be converged to any degree desired and to be cut into layers at will (STL formulation), or at defined points cuts are made directly into the CAD model (contour-orientated or SLC formulation). New CAD models support this procedure. The first mentioned triangulation procedure leads to so-called STL data records, and the direct method to contour or SLC data records.

The definition of the model surface by triangles and the reproduction of this definition in the form of an STL data record represents de facto an industrial standard. That these auxiliary means have been known for a long time, for example, in connection with shading 3D components, has helped. The decisive element, however, in establishing the STL format as stereolithography or rapid prototyping interface has been the fact that interface formulation was already published at an early stage and could therefore be used by all producers, suppliers of CAD systems, and especially software houses concerned with the formulation of special rapid prototyping software (support generator, etc.) irrespective of the supplier.

STL Format

Based on the definition of an entire solid surface by means of (small) triangles, a mathematically defined cross section can be made through the body for every arbitrary z-coordinate and its contour can be calculated. The convergence of the mathematically exact contour (as provided by the CAD) by triangles is an inaccuracy by any measure and is larger the lower the number of triangles chosen. In Figure 2-10a, this fundamental secant error is demonstrated in the example of errors appearing in the convergence of a circle by (f/4), eight (f/8), and twelve (f/12) secants. Figure 2-10b shows the consequences on the modeling of a globe surface.

With the increased accuracy in defining the surface made possible by increasing the number of triangles, the amount of data increases enormously. Critics continually cite this as one of the disadvantages of STL formulation. Although this is correct in principle, it should be remembered that, if alternative processes are used, for example, the contour-orientated formulation, the closed curves too must be displayed as polygonal drawings and that the amount of data resulting from this kind of representation also grows enormously with the growing demand on accuracy.

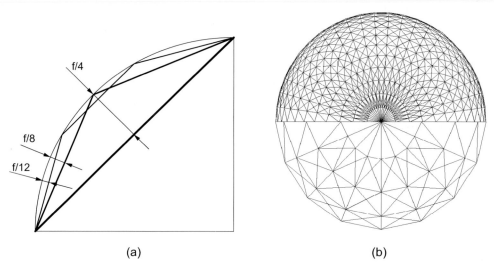

(a)   (b)

Figure 2-10   (a) Secant error in the convergence of a circle by 4 (f/4), 8 (f/8) or 12 secants (f/12)
(b) Influence of the number of triangles on the modeling of a sphere (STL)

Nevertheless, it should not be overlooked that good convergences of STL formulations will cause the amount of data for a free-form surface model to grow considerably. In one published example the growth factor was 22 [DON91].

The STL formulation, however, possesses practical advantages:

- Because the interface contains only data elements of a type that can be described by relatively simple means, syntax errors in the programming of the interface are very easy to recognize and to eliminate, and therefore pose practically no problem.

- In contrast to contour orientated interfaces smaller errors may be repaired relatively easily. It is also an advantage that a triangle provides a higher quality of geometric information than that of the contour vector.

- STL models permit any desired scale at random without reversing into the CAD.

SLC Format

Although attractive from its theoretical approach – but in practice development has only started – the SLC-format – or in general all contour-orientated formulations – assume that they are able to define directly the edge of any arbitrary slice through a body. Following the example of established plotter software, this can also be done, for example, on the basis of HPGL files (Hewlett Packard Graphic Language) or similar standard formulations. Well known are:

SLC, SLI   3D-Systems

CLI   EOS

HPGL   HP

SLC   Stratasys

F&S   Fockele und Schwarze

The contour is then derived directly from the CAD which is the mathematically exact object [FOC94]. In any case one has to keep in mind, that these formats are entirely 2D formats.

The information from the 3D CAD is sufficient to specify hatchings directly in addition to the contour, thereby describing the contour completely. External and internal edges and hatchings of the cross-sectional areas are differentiated (Figure 2-11).

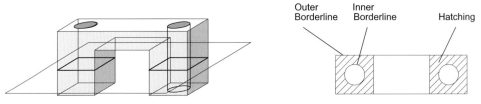

Figure 2-11   Direct contour generation from a 3D CAD

In practice a grave disadvantage of the SLC formulation is that data records cannot be scaled later in the CAD without being further processed, as only layer information without z-relation is available. STL data, on the other hand, define the entire surface and are easily scaled.

Direct slicing in CAD requires, of course, that either the complete CAD model is available to the model maker for the creation of the rapid prototyping model (program and current data record), or that the technical designer himself must complete this part of the model maker's work.

The adoption of sliced contour data leaves the model maker no possibility of influencing the creation of the geometrical base for the rapid prototyping process with the aim of optimizing the model.

The alternative transfer of the entire CAD data requires that the model maker himself uses that particular CAD system, and has a good command of it, which for more than two systems must normally be ruled out for cost reasons alone. That only leaves the possibility of transferring the data to a second CAD system, which is usually unacceptable owing to the high risk of error.

Very often security reasons are cited for not adopting the entire CAD data record in the aforementioned early phase of development, for, clearly, more information is passed on than that of a pure STL geometry formulation.

A solution could be the more recently available rapid prototyping software which has the character of surface modeling CAD programs, reduced in their functionality and which shows restricted but still sufficient functionality (see also Section 3.1).

In any case and irrespective of the formulation of the interface it should be borne in mind that by the (mathematical) generation of the layers to be built (sliced) and irrespective of their formulation, the continuous 3D model is transformed into a model stepped in the z-direction. The usual layer thickness of about 0.1 mm creates the effect of clearly recognizable stair stepping effects with corresponding accuracy and surface qualities depending on the production direction. Whereas in one layer, that is in x-y direction, accuracy of 0.05 mm can be achieved – depending on the kind of contouring (scan strategy, etc.) and the rapid prototyping process – they are therefore distinctly worse in z-direction (exception: prototyper for precision parts and micro components, see Section 3.3.1.6).

Figure 2-12 shows this stair stepping effect in the example of a drilling parallel to the direction of the layers. For further clarification, the situations of a drilling vertical to the layer direction are shown for comparison. It is assumed that the circular contours in the layers are generated continuously. In Figure 2-13 this stair stepping effect is clearly discernible on a stereolithography component.

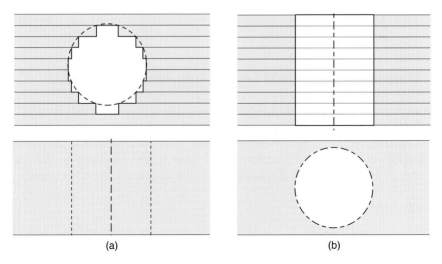

(a)   (b)

Figure 2-12 Principle-governed stair stepping in layer-orientated processes using the example of a drilling (a) parallel to the layer direction and (b) vertical to the layer direction

Figure 2-13   Principle-governed stair stepping on a stereolithography component (layer thickness: 0.125 mm). Photo: CP

2.2.3   Projection of Geometrical Layer Information on the Layer

The accuracy with which the mathematical geometry data are converted into physical layers basically depends on the contouring process used. Should a contouring process (exposure strategy: scanner, plotter, lamps or extruder, print head, etc.) be chosen the subsequent model-making results are largely predetermined. This applies especially to those widely used processes that use scanners, that is, processes that are primarily laser-supported.

Laser-Supported Contouring

The laser is of special importance in prototyping due to its ability to focus high energy density on to an extremely small operating diameter. Further details of the laser-principle and it's application for material processing can be found in [MIG96] and of the types of laser preferred in prototyping in appendix A2-3.

We differentiate between three processes when projecting the geometrical layer information onto the layer: vector, raster, and mask processes.

In the *vector process,* the single contour elements are generated continuously from basic geometric elements such as straight lines, circular arcs, and so forth on the basis of the standardized plotter software. It can be assumed, therefore, that a circular contour parallel to the layer will appear as a polygonal curve in the STL formulation and as a continuous circle in the SLC formulation.

In the *raster process* the contour is still generated using the same geometric information, but – as with a television picture – line by line. As a result, a stair stepping effect appears from line to line similar to that in the z-direction. The height of the steps is determined by the width of the effective track, and, in the case of a laser, by the width of the laser beam. As the effective width of the laser beam may be up to 0.3 mm, such raster processes will, for non-rectangular geometries, have a much higher tendency to generate a stair stepping effect in the layer surface than that of the design-dependent layers in the z-surface. The vector process enables the contour to be moved by half a beam diameter (beam width compensation), resulting in very accurate contourings; the raster process cannot do this.

The third possibility of reproducing the geometric information of a single layer on the layer is the mask process. Here a mask is produced geometrically similar but on a smaller scale, which is then screened by an energy source. As with a diapositive an authentic-scale contour is reproduced on the layer surface. This process works relatively quickly and is especially independent of the complexity of the layer information. The accuracy is limited by the accuracy with which the mask can be produced.

The vector process is in principal the most accurate process, but at the same time the slowest. The time required for the generation of a layer depends on the complexity of this layer.

Figure 2-14 shows the three principle processes for the reproduction of the geometric layer information on to the layer.

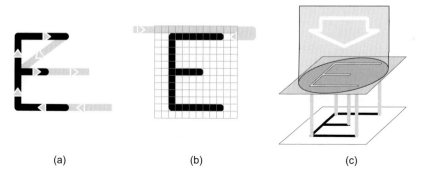

| (a) | (b) | (c) |

Figure 2-14 (a) Vector process. (b) Raster process. (c) Mask process for the projection of geometric layer information onto the layer

Contouring Without a Laser

There are a growing number of processes, especially newer processes, that do not employ a laser for contouring. Extrusion processes usually use heated nozzles through which material is continually deposited, while ballistic processes and 3D plotters preferably fall back on piezoelectric or bubble-jet print heads such as those often found in computer printers. Finally, cutter plotters fitted with knives backed up by cutter laser are used for contouring in layer-lamination processes.

Such contourings – apart from the characteristics of the really contouring element, for example, the nozzle diameter – all have the droplet diameter and the blade diameter in common which primarily determine the precision of the employed x-y-(z) handling system and the quality of the model. The mechanical setup, the incremental actors, and the quality of the control are important influencing variables.

2.3 Generation of the Physical Layer Model

All rapid prototyping processes work in two steps when generating the physical layer model:

- Generation of a cross section (x-y plane)
- Joining this layer with the preceding one (before, during, or after the layer generation, z-direction)

Joining the layers in the z-direction with one another is achieved in the same way as joining them in the x-y direction, with the exception of layer lamination processes: The energy or the amount of binder necessary for the joining is proportioned in such a way that not only the layer itself but also part of the preceding layer is affected and thus joined to the new layer. In layer lamination processes the layers are cut out of foils with predetermined thicknesses (z-increment) and glued on top of each other.

For the implementation of the rapid prototyping principle several fundamentally different physical processes are suitable:

- Solidification of liquid materials (polymerization process) (Section 2.3.1)
- Generation from the solid phase (Section 2.3.2)
  - Cutting from foils or ribbons, milling from slabs or other materials (layer process)
  - Incipiently or completely melted solid materials, powders, or powder mixtures (extrusion and sinter processes)
  - Conglutination of granules or powders by additional binders
- Generation from a pasty phase
- Precipitation from the gaseous phase (Section 2.3.3)

2.3.1 Solidification of Liquid Materials Photopolymerization – Stereolithography (SL)

All processes in which the underlying mechanism is solidification of liquids (including pasts) are based on the concept of photopolymerization. They use a viscous monomer with few or no crosslinks that is interspersed with suitable photoinhibitors. Exposure to ultraviolet radiation sets off a spontaneous polymerization, in the course of which the liquid monomer becomes a solid polymer. This process, which in principle also works under sunlight is adjusted to the special requirements of rapid prototyping regarding exposure strategy. Either a fine laser beam describes the contour of the particular cross section onto the surface of a resin bath (and that is the most common method), thereby generating the necessary critical energy density and consequently the necessary solidification locally, or the entire cross section is imaged onto a transparent mask by a process similar to photo-copying and is projected by means of strong UV lamps through this mask onto the surface of the resin bath. There are also processes that trigger the polymerization thermally. Today, however, they are still of no commercial significance.

Basic Principles of Polymerization

Polymerization is a chain reaction in which unsaturated molecules are linked to macromole-cules (polymers) [GRU93]. The liquid mixture of single molecules (monomers) is converted into crosslinked, cured plastics.

The monomers consist mainly of hydrocarbon compounds and possess double linkages that can be broken by heat or catalysts. A chain reaction follows in the course of which the monomers are linked to polymer chains. Depending on the type of reactive particles bringing the polymerization about, so-called radical, cation, and anion polymerizations are distinguishable. For the use in rapid prototyping processes (see Section 3.3.1) the radical and the cationic polymerizations are relevant. There are three different phases in polymeri-zation:

- Initial reaction
- Propagation reaction
- Termination reaction

The solidification of acrylates is based on *radical polymerization.*

During the initial reaction, an external energy input causes the initiators contained in the resin to disintegrate into their radicals, which are then left with a single unpaired electron on their outer shell [GRU93]. As the particle seeks to find a partner for its electron, radicals are extremely reactive.

During the propagation or growth reaction, the radicals react with the double-linked hydro-carbon in the monomer. Since the double linkage is broken, the monomer is also acquiring on one side a single electron on its outer shell. The new group reacts further with other monomers forming long polymer chains. The chain reaction is brought to an end by the termination reaction. The termination is caused by:

- Interlocking of two polymer chains into one
- Reaction of the polymer chain with the radical of an initiator
- Transfer of a hydrogen atom to a macroradical
- Elimination of a bondable hydrogen atom

The linkage of monomers to polymers is explained in the example of the polymerization of ethylene to polyethylene (Figure 2-15).

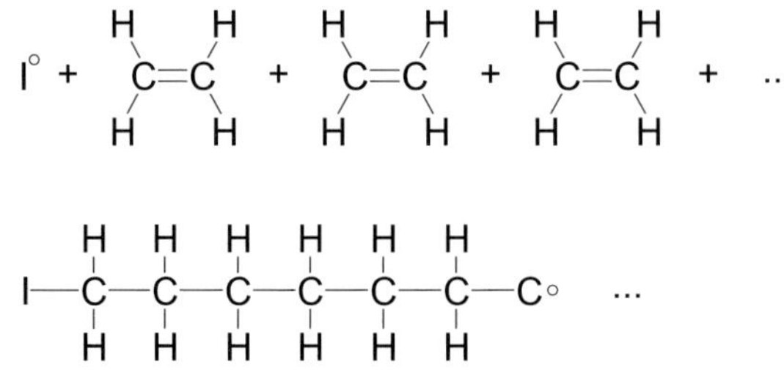

Figure 2-15 Polymerization of ethylene to polyethylene

The solidification of epoxy resins and vinyl ether resins is based on *cationic polymerization*.

In cationic polymerization, positive-charged ions (cations) always are found at the end of a chain. During the initial reaction, cations are generated from a catalyst. In most cases the catalysts are acids.

During the propagation reaction, ionic linkages are formed between the cation and the monomer. Further monomers can then be added to the end of the chain, which is now cationic.

The termination of the chain is brought about by an addition of anion from the catalyst. The termination of a chain by linkage of two polymer chains is impossible, as all reactive chain ends are positively charged.

An example of a cationic polymerization is the polymerization of isobutylene to polyisobutylene (Figure 2-16).

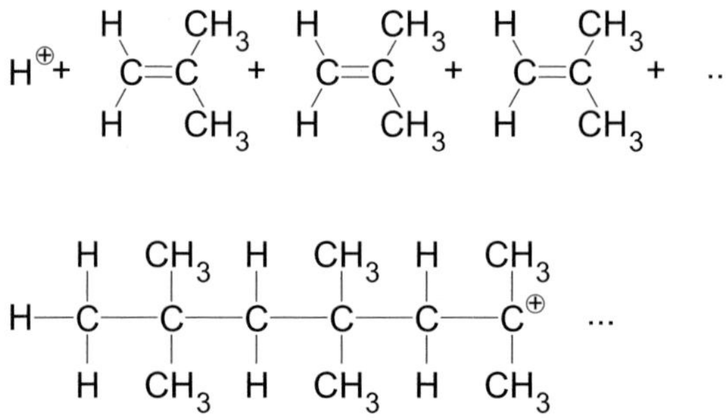

Figure 2-16 Polymerization of isobutylene to polyisobutylene

In stereolithography it is necessary to limit polymerization to allow specific single areas to be cured. In contrast to the polymerization processes in large-scale industry in which the disintegration of catalysts, for example, is started by heat input, the method employed is different here. The monomer is mixed with an initiator that disintegrates into two radicals or ions via the action of a photon of a certain wavelength. Because the polymerization can be started only with the aid of radicals or ions it is limited to areas affected by photon radiation.

The degree of polymerization is influenced by several factors. Among these is the presence of a degradative reaction, opposed to the growth reaction, that causes the monomers to separate from the polymer chains. The speed of this degradative reaction increases as the temperature increases so that at a certain ceiling temperature growth and decomposition are in equilibrium. Because polymerization reactions are highly exothermic (50 to 100 kJ mol^{-1}), large polymerization installations have to be cooled continuously to keep them under the ceiling temperature. It is not necessary to cool the resin in rapid prototyping installations as the polymerized amounts are too small. To achieve a temperature of approx. 25 °C to 30 °C which is beneficial for the flowability of the resin, even the resin container needs to be heated depending on the temperature of the environment.

The circumstances can change in the future with improved laser performance.

In addition to the degree of polymerization it is necessary to take into account the so-called polymerization rate, which defines the ratio of monomers not participating in the polymerization to the total mass. In contrast to the polymerization degree the polymerization rate grows with increased viscosity.

When monomers are linked to polymer chains the material is densified and the resin shrinks. This shrinkage causes tensions in the consolidated resin which can lead to warping and cracking. The extent of volume shrinkage for epoxy resins is 2% to 3%; for acrylic resins it is 5% to 7% (see also Appendix A2-15 to A2-19).

Laser-Induced Polymerization

A locally limited polymerization can be started by one of two processes:

- Reproduction of the entire layer by a smaller scale mask with the aid of a powerful ultra-violet light source (mask process, Figure 2-14c).

- Exposure to an ultraviolet laser beam, which "writes" the desired contours into the resin surface by means of certain scanning strategies (vector or raster processes, Figure 2-14a, b).

Although there are also processes that work with masks, laser stereolithography is the most important with respect to industrial implementation (see also Chapter 3). The basic principles of the laser and the characteristics of laser radiation are explained in [MIG96].

In the following passage some specialities are discussed which result from the use of laser-radiation sources for photo-polymerization.

Depth of the Cure Track

The local degree of polymerization and the rate of polymerization depend on the number of photons that pass through a certain activation cross section of the resin, thereby potentially reacting with the initiators.

From a critical surface energy (critical energy) onward so many photons react with the resin that it is transformed from a liquid to a solid state. This transformation point is called the gel point. At first the resin does not have any mechanical stability; only after the surface energy is increased the resin sufficiently polymerized to carry mechanical strain. The boundary surface inside the resin between the solid and the liquid state is formed by that surface in which the surface energy exactly corresponds with the critical energy.

For the surface energy irradiated on average onto the resin surface (z = 0), which is also the maximum energy affecting the resin surface, the following applies:

$$E_{max} = \frac{P_L}{v_s \cdot h_s}$$
(2.1)

with E_{max} = surface energy on the surface at z = 0
P_L = average laser-performance
v_s = speed of the laser beam
h_s = hatch width

The absorption within the resin follows the Beer-Lambert equation:

$$E(z) = E_{max} \cdot \exp\left(-\frac{z}{D_p}\right)$$
(2.2)

with $E(z)$ = surface energy in the depth z
E_{max} = surface energy on the surface at z = 0
D_p = optical penetration depth of the resin.

The optical penetration depth D_p of a material is defined as the path length after which the intensity of a transmitted beam has dropped to the 1/e-fold part, or its energy to the $1/e^2$-fold part.

With the aid of the Beer-Lambert equation and the definition of critical energy E_c the cure depth C_d, down to which the resin is cured, is given as.

$$C_d = z(E_c) = D_p \cdot \ln\left(\frac{E_{max}}{E_c}\right)$$
(2.3)

With the first relationship between the laser performance and energy on the surface it follows:

$$C_d = D_p \cdot \ln \left(\frac{P_L}{v_s \cdot h_s \cdot E_c} \right)$$
(2.3 a)

As C_d is proportional to the logarithm of E_{max}, a straight line results if logarithmic representation is employed. The gradient of the straight line is defined by the value of D_p. The critical energy density, at which the cure depth is zero, is defined by the intersection of the straight line with the abscisse. As the curve is dependent only on the resin constants E_c and D_p and the maximum energy density, it is characteristic for a certain kind of resin and is also called a working curve (Figure 2-17) [HEL 95][1].

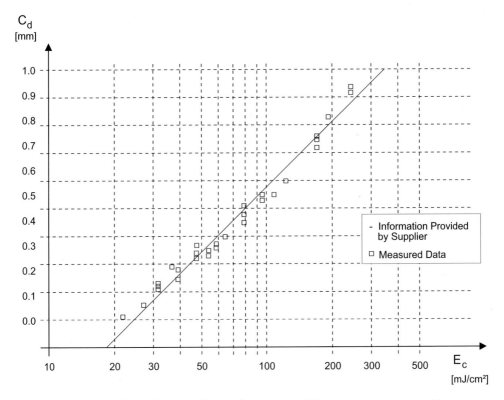

Figure 2-17   Cure depth as a function of the surface energy and the resin parameters, working curve for the resin HS 660

Given the resin parameters E_c, D_p, and a constant laser performance P_L the cure depth of the resin can be established on the basis of the speed of the laser.

---

[1] Today the resin HS 660 is no longer of any technical importance. However, Figure 2-17 clarifies very well the principal relationships, which are also analogously valid for modern resins.

Contour of the cure track

In addition to the cure depth, the width and the shape of the cured track are significant. This calculation cannot be based on an average surface energy; the energy distribution of the Gaussian beam must be taken into account. For this purpose a system of coordinates is fixed with its x-axis in the direction of the laser speed vector and its z-axis in the direction of the beam. The surface fixed by the x- and the y-axis coincides with the resin surface. A cutting plane is laid through the zero point onto which the contour of the cure track is to be calculated. The distance between the midpoint of the laser beam and a point Q on the cutting plane is called r (Figure 2-18).

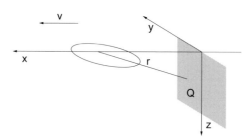

Figure 2-18 System of coordinates for the calculation of track geometry

For the intensity distribution in the Gaussian beam and with the relation between the intensity in the beam midpoint I_0 and the laser performance P_L at a Gaussian beam.

$$I_0 = \frac{2 \cdot P_L}{\pi \cdot \omega_0^{\,2}} \tag{2.4}$$

with ω_0 = beam radius on the resin surface.

It follows:

$$I(r,0) = \frac{2 \cdot P_L}{\pi \cdot \omega_0^{\,2}} \cdot \exp\left(\frac{-2 \cdot r^2}{\omega_0^{\,2}}\right) \tag{2.5}$$

The surface energy at a specific point on the section plane (y, z = 0), over which a laser beam runs with the intensity I(r,0), equals the temporal integration over intensity. By replacing I_0 according to Equ. (2.4) the surface energy on the resin surface results.

The periphery is to be found exactly where E(y, z) equals the critical energy E_c. By equating $E(y^*, z^*) = E_c$ and transforming, we arrive at the following equation:

$$\left(\frac{2}{\omega_0^{\,2}}\right) \cdot y^{*2} + \left(\frac{1}{D_p}\right) \cdot z^* = \ln\left[\sqrt{\frac{2}{\pi}} \cdot \left(\frac{P_L}{\omega_0 \cdot v_s \cdot E_c}\right)\right] \tag{2.6}$$

Equation (2.6) is the definition of a parabola. The cure track therefore has the geometry of a parabolic cylinder as shown in Figure 2-19.

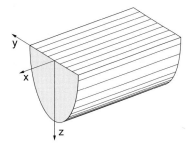

Figure 2-19 Parabolic shape of the cure track under the influence of a Gaussian-beam on a photopolymer

From Equ. (2.6) the relations for the track width L_W and the cure depth C_d can also be derived [JAC92], [FAS94].

Optimization of the Layer Thickness

Knowing the geometry of the cure track enables a layer thickness to be determined in which the time needed for curing a certain volume is minimal. For a reasonable coverage of the volume with parabolic cure tracks it is assumed that the cure track has a distance of 0.02 mm to both the lower layer and the adjoining layer.

With d as layer thickness for the curing time of a certain volume, the following proportion-alty results:

$$T \sim \frac{1}{v_s \cdot d \cdot \left(L_w + 0.02mm\right)} \tag{2.7}$$

With the cure depth $C_d = d - 0.02$ mm and insertion of the relation for the track width L_W it follows:

$$T \sim \left\{ d \cdot \sqrt{\frac{2}{\pi}} \cdot \frac{P_L}{\omega_0 \cdot E_c} \cdot \exp\left(-\frac{d - 0.02mm}{D_p}\right) \cdot \left[\frac{\omega_0}{\sqrt{2}} \cdot \sqrt{\frac{d - 0.02mm}{D_p}} + 0.02mm\right] \right\}^{-1} \tag{2.8}$$

Equation (2.8) is shown graphically in Figure 2-20. The minimum of the function is d = 0.3706 mm for an assumed optical penetration depth $D_p = 0.25$ mm. As only proportionali-ties were investigated there are no resulting absolute values for the ordinate.

The penetration depth D_p of 0.2 to 0.3 mm of different resins lies within the technically sensible dimension of 0.1 to 0.5 mm. If deeper layers are required, the surface energy needs to be increased, which simultaneously lowers the scan speed. Therefore, attempts are being made to produce resins that have a greater penetration depth. As shown in the following subsection, the alteration of the penetration depth also influences the stability of the compo-nents.

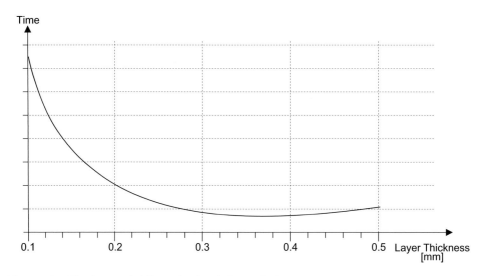

Figure 2-20 The time needed for curing in relation to the layer

If resins with a greater optical penetration depth are used it should be taken into account – especially in combination with higher powered lasers – that leakage radiation can result far more easily in undesired polymerization than previously and thereby have a negative influence on the accuracy of the components and above all on the aging of the resin.

Effect of the penetration depth on the stability of the component

By increasing the penetration depth D_p of the resin with a constant cure depth C_d the required surface energy on the surface of the resin is lowered. Accordingly, fewer photons are absorbed and the rate of polymerization decreases, which results in less stability of the component. The relevant value excess energy E_x is introduced here. This is a measure for the amount of energy available for the polymerization in addition to E_c. A useful definition for E_x is given in Equ. (2.9) [JAC92].

$$E_x = \left(\frac{1}{C_d}\right) \cdot \int_0^{C_d} \left(E(z) - E_c\right) \cdot dz \qquad (2.9)$$

After implementing Equ. (2.2) it follows that:

$$\frac{E_x}{E_c} = \left(\frac{D_p}{C_d}\right) \cdot \left[\exp\left(\frac{C_d}{D_p}\right) - 1\right] - 1 \qquad (2.10)$$

This function is illustrated graphically in Fig. 2-21.

The following fundamental relationships derive from Equ. (2.10) and Figure 2-21:

- The excess energy is directly proportional to the critical energy E_c. A reduction of E_c to achieve shorter build times will result in a direct decrease of stability during the green phase.

- By reducing the penetration depth or raising the layer thickness, the stability in the green state is increased. The effect is especially prominent with values of $C_d/D_p > 3$.

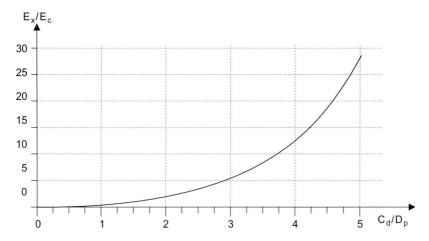

Figure 2-21 Excess energy relative to the cure depth and the optical penetration depth

The fundamental relationships discussed are valid for all photopolymerization processes, especially for laser-supported stereolithography processes. This explains the fundamental advantages and disadvantages of stereolithography processes which – with process specific restrictions – are in principle valid for all industrial processes discussed later.

Advantages of Stereolithography

Stereolithography, also known as stereography, is at present the most accurate of all rapid prototyping construction processes. Its accuracy is limited by the machine, but not by physical limits. For example, the minimal depictable land widths are in principle a function of the laser beam diameter. The tenuity of the z-stepping is not limited by the process. It is limited by the wettability of a solid layer by the (following) liquid monomer layer, expressed as the relationship of volume force (proportional to the layer thickness) and the surface tension. Thin layers consequently tend to "rip." These physical obstacles can be overcome if the layer thickness is limited by solids such as glass plates rather than a free surface.

It is in principle possible to contour the boundary of the x-y planes in the z-direction by appropriate control (five-axis) and exposure strategies (variation of pulse-pause relationship and laser performance) and thus to achieve a quasi continual z-modeling.

Stereolithography not only allows the production of internal hollow spaces, as do nearly all the other rapid prototyping processes, but also permits their complete evacuation as a result of the process technology. For this a drainage opening is necessary which should be clearly much smaller than the diameter of the hollow space. Together with the further advantage of the materials being transparent or opaque, this facilitates the visual judgment of internal hollow spaces as is, for example, necessary for medicinal use (e.g., mandible, mandibular nerve channel).

Complex models, or those of larger dimensions than the build chamber, can be assembled from single partial models into arbitrary complex complete models. If the same photosensitive resin is used as binder and UV radiation sources for local curing, the section points are unnoticeable in respect to their mechanical-technological properties and they are also invisible to the eye.

The models can be finished by sand blasting and polishing and, to a certain extent, by machining and coating.

Noncrosslinked monomers can be reused, and completely polymerized resin can be treated as household garbage.

Disadvantages of Stereolithography

Owing to its process technology, stereolithography is restricted to photosensitive material. When developing resins, therefore, this property is the most important. The usual primary properties such as resistance to extension, elasticity, temperature stability, and so forth are of secondary importance. Furthermore, material development is limited to stereolithographic usage and in view of the costs apportionable to the product it is correlated only with this market.

Stereolithography is in principle a two-stepped process in which the models are first solidified to a high percentage (> 95%) in the actual stereolithography machine; afterwards the finished model is placed into an oven to build up further crosslinkages until it is cured completely (this does not apply to printing processes or mask processes, SGC).

The green product must be cleaned with solvents (TMP, isopropanol). This requires the storage, handling, and disposal of solvents and is another time-consuming process.When making stereolithography models unsupported structures and certain critical angles of overlapping model parts cannot be realized without support, as during its generation in the resin bath the model is still a relatively soft green product. On the one hand these supportive structures need to be fitted when the model making is in preparation, and on the other hand they have to be removed manually from the green product or from the cured model.

To a small extent, photosensitive acrylates absorb oxygen, whereas epoxy resins are hygroscopic; this has to be taken into account when storing and processing the material. The models tend to creep even after being completely cured. After a few days or weeks unsupported walls show saggings that disappear if the model is turned over or supported. The newest epoxy resins show these characteristics less prominently.

2.3.2   Generation from the Solid Phase

2.3.2.1 Melting and Solidification of Powders and Granules – Laser Sintering (LS)

Powders or granules arranged in a powder bed are source materials for the formation of a solid layer whereby layer surfaces are melted together by a laser beam. Such processes are called sinter processes because the deliberate melting resembles a classical diffusion-controlled sinter process.

The term "sintering" derives from powder metallurgy and describes a procedure in which powdered material under high temperature and high pressure over a relatively long period of time is "baked" in a mold into a solid shape. The tabletting as well as the long processing time are characteristic dimensions. Sinter processes are used in powder metallurgy when:

- The high temperature of the molten mass makes other production processes uneconomical (Wolfram filaments of bulbs)

- A certain porosity is required (dry running properties of friction bearings)

- Alloy constituents are to be realized that in the molten state would segregate completely or partly (pseudoalloys)

Selective Laser Sintering

In the classical understanding of sintering under high temperature and high pressure two neighboring particles are linked to one another at a contact point first in the form of a neck, which is formed by the mechanism of surface diffusion. With the progression of sintering, over a longer period of time under the combined influence of temperature and pressure, material is transported especially along the particle boundaries and inside the sintering particles (particle boundary or volume diffusion)

The laser sintering used as a rapid prototyping process functions to the exclusion of the two fundamental components of the classical sinter process, pressure and time. Only a short thermal activation of the particles to be sintered takes place.

The particles lie loosely in the powder bed. They are made molten on the surface locally (selective) by a laser beam and thereby joined to a layer. In fact this is a partial melting and solidification process known as liquid sintering or (selective) laser sintering.

Because the classical conditions of high pressure and long contact time are, as previously mentioned, not required for laser sintering, it must be assumed that the laser sintering process is not or is not dominantly diffusion controlled, but that it is in fact an incipient melting or even a fusion of powder particles.

The laser sintering process exploits the fact that powder has a greater surface area than solids and that every physical system strives to minimize its energy state. Consequently in neighboring particles that are incipiently molten on the surface the total surface is minimized by fusing of the particles' outer skins.

The sinter process is best described as the interaction between the viscosity of the incipiently molten particle areas and their surface tensions. Both the counter effects are functions of the temperature and the material.

Materials for Selective Laser Sintering

In principle, all materials that can be melted and, after cooling, solidify again (as far as possible with a constant volume) can be used for the sinter process. Because – in contrast to the classical sinter process – the parameter pressure and temperature have to be kept very low only plastic material and wax were sintered at first mainly owing to their low melting points.

Plastics Powder

The low temperature range of up to approx. 200 °C favors the sintering of plastic materials. The low heat conductibility is also advantageous. It helps to limit the melting bath locally and to prevent the "growth" of models by neighboring particles being sintered on. Although the surface tension of the molten mass – low compared to that of metals – impedes the sinter process it fosters the wetting of the model, thereby minimizing the danger that macroscopic globules are formed and separated from the layer.

Crystalline and amorphous plastics behave completely differently. The vast majority of materials (metals included) solidify crystalline. Their elementary particles, atoms or molecules, are set in defined regular spatial crystal gratings.

Glass, resin, or pitch solidify amorphous (i.e., with no fixed structure). Although their elementary particles are also practically stationary they possess no recognizable defined structure. Physically they should be considered as an undercooled molten mass. The effect of devitrification is well known: an unwanted crystallization causes amorphous transparent glass to crystallize and become opaque.

Amorphous plastics are characterized by a broad temperature range within which they soften or convert into another state of aggregation without sudden alterations to their mechanical-technological properties (Figure 2-22a). Crystalline plastics change their state of aggregation and thereby all important mechanical-technological properties in such a narrow temperature range (often only 1/10 °C) that it is called a "melting point" (Figure 2-22b).

The basic properties of these materials are decisive for the behavior of models during the build process (shrinkage) and for the achievable properties (solidity) of models.

Polyamide (nylon) belong to the most often used sintering materials which are crystalline whereas polycarbonate and polystyrene are amorphous plastics.

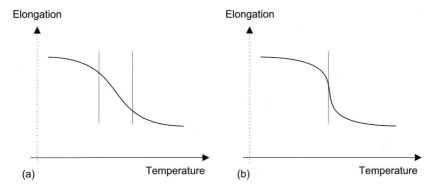

Figure 2-22   Physical properties of (a) amorphous and (b) a crystalline plastics, schematic

Multicomponent Metal-Polymer Powder

The first step to sintering metal powders is made via polymeric-linked metal powder. The process is fundamentally therefore not very different from the fusion of plastic powders. The polymeric shells are sintered on and "glue" the integrated metal particles together to form a so-called green product.

In a second step, the polymeric particles are expelled by heating while in a third step they are infiltrated with a low-melting metal (e.g., copper) in an oven (with reduced atmosphere). This process makes high demands on technology and installations.

From the viewpoint of sinter theory the process belongs to the category of plastic sinter processes.

Basically, all materials can be processed in the same manner in plastic sinter machines. Polymeric-linked sands, ceramics, and the above-described metals are technically realized.

Multicomponent Metal-Metal Powder

When low-melting metal powder – instead of polymeric binder – is mixed mechanically with the high-melting component, the process becomes safer and faster. The binder function is then taken over by the low-melting component. The mechanical-technical properties of the model, however, are very different from those of currently used series alloys. These models can therefore be used only to a limited extent as functional models. Figure 2-23 shows the basic interrelationships of polymeric linked powders and of multicomponent powders in the sintering process.

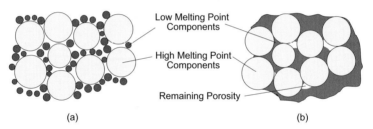

Figure 2-23 The principle of liquid sintering (FHG-ILT)

Single Component Metal Powders

At first the sinter process was used for plastic materials owing to relatively low process temperatures and poor heat conduction, which locally limits the reaction zone; now, with a modified process control, metals are also suited for laser sintering. In their melted state they have a higher surface tension and a lower viscosity than plastic materials. The driving force hereby is the surface tension while the opposing force is the viscosity, both of which are more favorable than with plastic materials. However, the process temperatures and therefore the constructive complexity are much higher. Such developments are the subject of worldwide intensive research, especially at the University of Texas (Austin), at the Fraunhofer-Institut für Lasertechnik (ILT) (Institute of Lasertechnology, Aachen, Germany) and Produktionstechnologie (IPT) (Institute of Production Technology, Aachen, Germany) and at the Fraunhofer-Institut für Werkstoff- und Strahltechnik (IWS) (Institute of Material and Beam Processing Technology, Dresden, Germany). In principle two paths are followed: the direct sintering of metallic materials according to the process of selective laser sintering described previously, and laser-supported generation derived from the classical coating process. The aim is the treatment not only of metals but also of ceramics. The characteristics of these processes, their current status, and the development trends are discussed in detail in Chapter 7. At the end of 1998, EOS presented a single-component metal powder for its machines (see Section 3.3.2.3) which produce metal prototypes with properties comparable to those of foundry steel.

Advantages of Laser Sintering

The linkage through thermal influence by laser beam gives selective laser sintering a far greater and possibly even an unlimited choice of materials in comparison to stereolithography. The resulting models are, depending on the material, mechanically and thermally resistant. In many cases they attain the status of functional models.

Powder that has not been sintered can be recovered and used again.

It is basically a one-step process. Further crosslinkage is not necessary. Supports are not needed.

For cleaning purposes (theoretically) only a brush and a sand blaster is needed. Solvents are not required.

The models are immediately ready for use.

Disadvantages of Laser Sintering

The achievable accuracy of the model is basically limited by the size of the powder particles. The material and its absorption properties and its heat conductibility define the possible build speed and the necessary laser power.

The models tend to "grow" depending on the relationship of "applied power" and "conductibility," that is, neighboring particles in the powder bed not belonging to the model are glued on by heat conduction and the model grows a "fur."

Internal hollow spaces are more difficult to clean than with stereolithography. It is unavoidable that loose or only slightly glued particles remain on the model. These particles impair accuracy and could possibly detach themselves at inconvenient moments when the model is later used. These components therefore cannot be sterilized for medical use in operating theaters.

To avoid oxidation, the sinter process takes place within an inertgas (nitrogen) atmosphere.

Because the sinter process takes place at near melting temperature, the entire powder bed needs to be preheated evenly to near this temperature to achieve an efficient sinter process. The temperature must be kept within restricted limits (a few degrees centigrade). The heating and cooling processes are very time-consuming.

2.3.2.2 Cutting from Foils – Layer Laminate Manufacturing (LLM)

The most simple method of producing 3D models is to split them into 2D contoured layers, to cut these layers out, and then to assemble them into 3D models. This is no longer a purely additive process. Bernard's precise term for all rapid prototyping processes which, in addition to the purely additive steps also include denuding (milling) or separating (cutting) steps, is subtractive/additive processes [BER98]. We will not follow this refined terminology, as the discussion here is focused primarily on the generation of models from layers; in practice scientifically exact, refined terminations do not enlighten further.

Layer models are known through the rib constructions of ships or the contour line models in geodesy. To obtain sufficiently detailed models, the layers must be very thin and exactly positioned to one another. Therefore, in most processes, the new layer is first glued to the preceding one, that is, onto the already finished part of the model; only then is it contoured. Layer model-making processes are known as layer laminate manufacturing (LLM). More common is the term laminated object manufacturing (LOM). This is not a generic term but a registered trade name which in this way became a synonym for the entire family of models (see also Chapter 3).

Lasers are very commonly used for contouring. Hot-wire cutters, knives, and milling machines are also used. Especially when contouring by milling machine the layers – in addition to being glued together – are also pinned together over center holes contained in the contour.

Advantages of Layer Laminate Processes (LLM)

LLM processes are especially suited for the generation of compact model parts without elaborate shading techniques and for large models. Layer processes work fast in relation to the model volume, and they are not very complicated from the machine-technical aspect. They work with practically any material. If paper is used, material costs are low and disposal is not a problem.

A continuous contouring of the x-y boundaries in the z-direction is, in principle, possible by means of a more complicated (five-axis) control of the laser or by milling.

The use of a laser for contouring the layer geometry has the advantage that – theoretically – an almost unlimited range of material is available. With the aid of the laser beam, all materials known today can be cut. By using correspondingly optimized laser cutting processes, cutting speeds well in excess of those presently realized are in principle possible (see Figure 2-24) [GEB95].

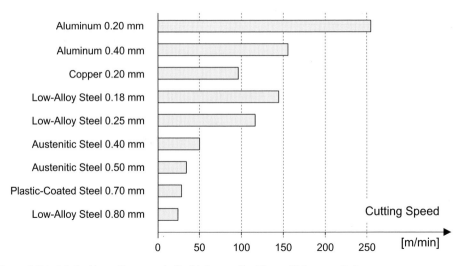

Figure 2-24 Attainable cutting speeds for high-speed cutting with laser-radiation

Disadvantages of Layer Laminate Processes (LLM)

All LLM process models have the disadvantage of having very different mechanical-techno-logical properties in layer direction and rectangular to the layer direction. It is also a disad-vantage that parts not required in the models such as internal cutouts either have to be removed manually from each layer or remain in the model as solid bodies that subsequently need removal (depending on the geometry) by means of complicated cleaning measures.

Slightly inclined slopes are realizable only with difficulty, as the superfluous material not belonging to the model cannot easily be removed. Paper models have to be impregnated after being unpacked. All in all final finishing costs are very considerable.

Most machines allow the use of only one material thickness, resulting in the production of models with only constant z-distances. However, there is no compulsive reason as to why the material that is available in several thicknesses should not be changed during the running build process.

The unused material is waste. For "shell character" models the ratio of model to waste can be as high as 1:9. Under these circumstances the favorable material price is put in perspective.

2.3.2.3 Melting and Solidification Out of the Solid Phase – Fused Layer Modeling (FLM)

Extrusion Processes

Solid wire-shaped materials of pasty or semimolten consistency are melted in a heated single- or a multi-nozzle system and deposited geometrically defined onto a structure. Due to loss of heat by conduction into the solid the physical model is formed by solidification. This process is outstandingly well suited for materials with a low melting point and low heat conduction. Theoretically there is no limit on the type of material used. In practice, however, metals and ceramics with high melting points present enormous problems with regard to the melting, the temperature gradient on the model, and also the required preheating of the workroom. As material feeding and material application are carried out simultaneously by the same nozzle, this process can be considered in principle as a fully adequate 3D process depending only on the control of the nozzle. Using single nozzle extruders, the relatively large material diameter limits the degree of detail achievable in the models. A suitable machine construction can mostly avoid these disadvantages (Section 3.3.4).

A satisfactory connection between the extruded material and the already finished part of the model is possible only when the material is "squeezed on," that is, making the diameter, which is still a circle in the nozzle, elliptic. Only then can the equilibrium between volume and surface tensions be regulated in such a way that the connection is as smooth as possible. However, it is essential that the longitudinal axis of the nozzle always remains on one particular plane, which forms a particular angle to the layer direction, depending on the process, and lies in z-direction.

Ballistic Processes /"Drop on Demand"

A variant of the melting of the solid phase and the subsequent solidification by cooling on the model is to "shoot" the molten material by means of a piezoelectric nozzle system or a bubble-jet print head onto the model, not continually but in the form of (spherical) droplets. Therefore the process also is called "drop on demand". The temperature of the material droplets thereby must be so dimensioned that, after application, they melt enough of the surrounding material to enable them to be firmly fastened to the model. The principal advantage thereby, in contrast to extrusion processes, is that no joints are visible and that a much higher degree of detail can be achieved depending only on the diameter of the droplets.

The process is not limited to thermoplastic material behavior but also works very well with any kind of sticky material as demonstrated by Objet's multi-nozzel stereolithography process (see Chapter 3.3.1.8.).

The droplets can impinge the material from almost any angle. Therefore this process is, in actuality and not only in principle, a fully adequate 3D process. It also permits in a further operation, theoretically at least, the application of material of various colors or other mechanical-technical properties as well as to polish surfaces locally or to improve them in some other way by applying other materials or different diameters of droplets.

Advantages of Extrusion Processes

Using single-nozzle extrusion processes enables relatively large volume amounts to be applied within a relatively short time. The resulting structures are solid. Model materials are used that are very similar or even identical to those series materials used in later production. The technical realization is comparatively simple. The material is completely used, leaving no waste. Solvents and similar agents are unnecessary. In contrast to most other rapid proto-typing processes, the application of various materials within one build process, or even one layer, is possible. The number of simultaneously applicable materials is limited only by the fact that the corresponding number of extrusion heads must be fitted geometrically into the machine and controlled by process techniques.

The process is realizable with machines that can be set up and operated in office environ-ments.

Disadvantages of Extrusion Processes

The main disadvantage of single-nozzle extrusion is that structures finer than the extrusion width cannot be produced. The same goes for details that, in the extreme case, may not be smaller than double the track width. This means that it is impossible to produce very fine grooves and especially fine ribs. The start of the extrusion always incurs a scab that, depending on the material, remains externally visible even after the contour is closed. Some materials tend to form filaments or to cause condensation.

Drop on demand systems are often very slow according to the number of nozzles per print head. The nozzles tend to clog up, requiring the installation of a suitable cleaning mecha-nism. Multi-nozzle heads (>100) with failure quotas for single nozzle openings in the %-region theoretically carry a high failure risk and are relatively expensive.

2.3.2.4 Conglutination of Granules and Binders – 3D Printing

The so-called binder processes, also known as 3D printing processes as they closely follow the 2D printing process of inkjet printers, are related to sinter processes albeit based on a different physical principle.

The starting point is again a bed of granules or powder. The powder particles are, however, glued together by an external binder that is injected according to the contours. The advantage is that the choice of material is theoretically unlimited; decisive are the character-istic properties of the binder.

Some process variants have the disadvantage that they involve two steps. The binder is expelled in a secondary process after which the model is resintered. This not only limits the choice of usable material but also creates the problem of scaling (shrinkage) which in general can be controlled reliably only by experience. A further negative point is that the realizable powder-binder combinations do not usually exist as series materials.

2.3.3 Solidification from the Gaseous Phase – Laser Chemical Vapor Deposition (LCVD)

The finest of structures are produced when a laser induces a chemical reaction in (N_2O) gas containing aluminum [$AlH_3 N (CH_3)_3$] and oxygen [LEH94]. The laser energy generated by two intersecting and focussing laser beams triggers a chemical reaction in the reaction zone that causes the formation of solid aluminum oxide. In this method, known as LCVD (Laser Chemical Vapor Deposition), aluminum oxide rods of 5 µm to 20 µm diameter are generated with a growth rate of up to 80 µm/s. An argon-ion laser is used. The application of an appropriate process control with a dimensionally-fixed reaction zone and a build platform that is moved at growth speed in the opposite direction to the growth direction enables the production of microfine structures without any supports. Figure 2-25 shows such a microstructure of aluminum oxide. In a further step the structure can be metallized completely or in parts and is then electrically conductive.

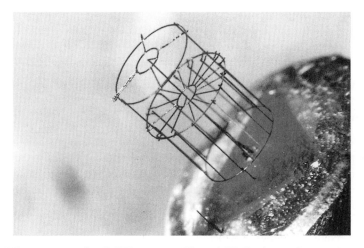

Figure 2-25   Microstructure of an LCVD process. (Photo: MPI-Göttingen, Germany)

The process allows the production of extremely filigree structures needed in microsystem technology and biomedical technology. It is also the basis for extremely improved surfaces. With this technique it is theoretically imaginable to metallize an arbitrary model and to improve its surfaces locally in an appropriate processing chamber. That the process is basically quite suitable for 3D processes and the fact that the sole precondition for producing the structure lies in its approachability for laser radiation supports this notion. It follows that either laser beams must be used that trigger the reaction and for which the rapid prototyping material should be permeable, or the process is restricted to surfaces approachable from the outside.

The LCVD process currently still in the trial phase is different from all other rapid prototyping processes in two basic points:

- It is a complete 3D process.

- It is not only a rapid prototyping process but also a real production process for microstructures for which no conventional alternatives exist.

2.3.4 Other Processes

In addition to the processes mentioned in the preceding, variants exist that until now could not be developed into a machine usable by industry. Polymerization is achieved by two laser beams intersecting at the point of polymerization [beam interference solidification (BIS)] or – as in industrial polymerization processes – by heat [thermal polymerization (TP)]. Another idea worth considering is triggering polymerization by the projection of holographic images into photosensible materials, that is, no longer in layers but along predetermined 3D curves [holographic interference solidification (HIS)].

Among the methods of gluing different foils together, it should in principle be possible to combine the foil and the polymerization process in such a way that different foils are glued together by polymerization [Solid Foil Polymerization (SFP)]. A detailed representation of such processes can be found in an albeit older but still excellent papers [KRU91].

Two other interesting ideas are based on lesser known physical effects.

Sonoluminescence

Sonoluminescence is the concentration of sound energy to such a degree that hollow spaces in liquids, especially air bubbles in water, radiate ultrashort light flashes in the UV range. It is at present not yet known for certain whether the light energy is strong enough to trigger a photopolymerization. Should this be the case this process would bring the great advantage that the UV radiation necessary for polymerization would not need to penetrate a material layer of a certain thickness and thereby reducing its strength; it could instead be generated directly at the required point of polymerization. Spatial structures could be constructed by such sound fields and their interferences.

Electroviscosity

Electroviscosity is the ability of certain materials to alter their viscosity within broad limits while under the influence of powerful electromagnetic fields. This can go so far that liquids completely solidify. The process is not yet in use in commercial industry, primarily because the increase in viscosity cannot be produced technically and economically over a longer period of time. However, this method opens new perspectives for rapid prototyping processes. Electroviscosity enables liquids to be solidified along a defined 3D curve that can be calculated using potential equations. They form thereby a continuous contour that can be filled in with photosensitive or other rapidly solidifying materials. This can be done within a relatively short time. After solidification, the contour is removed by liquefaction and on the basis of the new contour calculated from potential theoretical equations another part of the model can be generated. Such a process opens the possibility of working entirely 3D but it depends, at least according to current knowledge, on potential-theoretically calculated three-dimensional curves.

2.4 Classification of Generative Production Processes

The Professional literature contains a number of different representations concerning the systematy of rapid prototyping processes. Under the assumption that rapid prototyping processes are generative production processes, one representation is favored (Figure 2-26), which orientates itself to the German standard of production processes (DIN8580)[1] and especially to the German standard of archetyping (DIN8581). Rapid prototyping processes are classified therefore according to the state of aggregation of their original material.

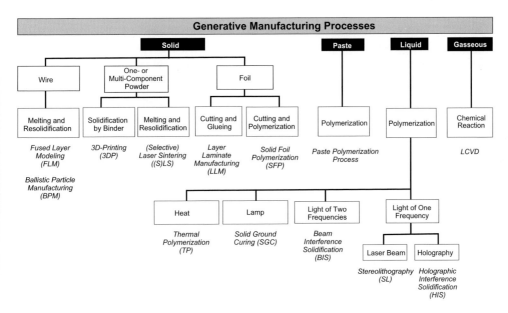

Figure 2-26 Classification of rapid prototyping processes according to the state of aggregation of their original material

---

[1] DIN Deutsches Institut für Normung, German Institute for Standardisation

2.5 Summarizing Evaluation of the Theoretical Potentials of Rapid Prototyping Processes

The evaluation of basic physical processes and their deducible advantages and disadvantages provide an insight into foreseeable development trends. The question of theoretical development limitations for single processes can at least be answered in the sense of a general trend. It is necessary to consider the limiting factors concerning the main points:

- Materials

- Model properties

- Details

- Accuracy

- Surface quality

- Development potential

The most important rapid prototyping processes, photopolymerization (stereolithography), melting of powders (laser sintering), cutting of foils (layer process), melting and solidification of wire materials (extrusion processes), ballistic processes, and injection of binders into powders (3D printer) are included.

Materials

A prerequisite for stereolithography is the presence of a photopolymer. Therefore, the useable materials will always be limited to photopolymers, that is, plastic material. This does not exclude all kinds of filling materials being used to modify these plastics.

The range of materials useable for sinter processes is theoretically limited only because these materials need thermoplastic properties in the widest sense. This means they must be meltable and after solidification they must be able to regain, almost at least, their former volumes and material properties. Of course, high-melting materials such as metals require a much higher complexity for their machine-technical realization.

The range of materials useable for layer processes is basically unlimited, as all currently known materials can be cut by a laser.

The range of materials useable for extrusion and ballistic processes is theoretically limited only because the nozzle material itself should not melt in the process. Here, too, the limitation is not so much theoretical but lies more in the technical realization.

3D printers have the widest range of material, as a large number of binder-powder mixtures is imaginable.

All processes operating with a bath or a powder bed need to contain more material than is necessary for the model, depending on the size of the build chamber. The filling material for a large stereolithography installation can easily cost about 50,000 $.

Model Properties

From the range of useable materials for polymerization, layer, and ballistic processes it can be seen that the mechanical stability of the resulting models is a limiting factor. In contrast, the stability of sinter models and extrusion models theoretically equals that of series materials.

Details

When assessing the ability to depict details a differentiation is made between the capability of describing thin wall or hollow spaces. With regard to wall thickness, the polymerization process is limited by the realizable laser beam diameter. It is to be born in mind, however, that polymerization processes are considered to be among the most accurate of processes and laser beam diameters of below 10 μm will be realizable in the near future (Filament laser).

In extrusion and ballistic processes the nozzle diameter and the resulting diameter of the extruded material define the minimum wall thickness. Hollow spaces, however, are not limited in any way in extrusion processes. The same applies to polymerization processes, because owing to the liquid material even delicate and spatially complicated hollow spaces can be emptied.

In sinter processes the minimum wall thickness and the internal hollow spaces are limited by the diameter of the powder particles.

In layer processes, in which the laser beam diameter is also important, the problem arises that very thin walls and internal hollow spaces are difficult to remove from the model.

Accuracy

In the polymerization process and the layer process accuracy in the build plane (x-y plane) is limited by the scanner-plotter unit. In the sinter process, on the other hand, the diameter of powder particles and in the extrusion and ballistic process the nozzle diameters and the resulting diameter of the applied material are the limiting factors.

The possibility of producing full 3D models is expressed by the ability to build continually in the z-direction and is in sinter processes also limited by the diameter of the powder particles. In other processes, however, acute angles present the limitation. In polymerization processes with free surfaces there is the additional problem that for very thin layers the wetting can no longer be guaranteed. In layer processes the ability to contour in the z-direction is limited by the fact that the ratio between the cut glue layer and the effective model-building material layer becomes too large. In addition there is a minimum foil thickness below which the transport of the foils can no longer be guaranteed. In the extrusion process it is basically the missing wettability again that limits the contouring in the z-direction. In ballistic processes, on the other hand, it is possible to work with micro-droplets.

Surface Quality

The qualities of surfaces are limited basically by the same factors as those affecting continual z-contouring.

Development Potential

In polymerization processes surface qualities and the degree of detail reproduction present the greatest potential for development considering that suitable copying processes can transpose the advantages of its extremely high accuracy to metal prototypes and series components.

In sinter processes the laser performance is limited by the realizable diameter of the powder particles and during the sintering by the ratio of powder particle diameter to laser beam diameter. (The powder particle diameter chosen cannot be smaller than a certain size.)

In layer and extrusion processes, especially the anisotropic model properties connected with this process, are considered a limit to the development. In extrusion processes also the relatively large diameters of their extrusion nozzles limit the realizable degree of detail reproduction.

Continuous 3D Model Generation

All currently available industrial systems are 2D processes with regard not only to their realization but also to their physical basic principle. They realize the shaping in one plane element and add the third dimension by stacking up neighboring planes. To differentiate between this process and a real 2D process in one plane they are referred to as 2½D processes.

Those processes that can also produce continuous contours in the z-direction by following physical basic principles at least, and superimpose them onto the x-y direction, that is, they are able to reach any spatial point, are called (real) 3D processes. There are two possibilities:

- The energy necessary for solidification is generated by superimposing two or more partial beams and by suitable control and interference any spatial point can be reached (beam interference).

- The material is transported by a moveable heated nozzle to any spatial point (five-axis CNC), where it is added to the model (by extruder or ballistically).

This potential of 3D capability is currently not yet technically applied (see Section 3.3.4.1). It is not merely a question of expenditures; the necessary software and suitable control algorithms are not yet available.

The qualitative properties described here are quantified in Appendix A2 and are attributed to materials and machines in the form of technical data.

3 Industrial Rapid Prototyping Systems

On the basis of the fundamental physical processes, in a little more than 15 years after the presentation of the first prototyper, more than 30 industrially useable prototypers for the direct computer-supported manufacturing of physical models have been developed.

The six classes of prototypers that currently represent the most important model-making processes are selected from a large number of available worldwide prototypers which will be discussed in detail in the subsections that follow:

- Stereolithography   (SL)
- (Selective) laser sintering   (SLS)
- Fused layer modeling   (FLM)[1]
- 3D printing   (3DP)
- Layer laminate manufacturing   (LLM)[1]
- Ballistic particle modeling   (BPM)

The technological generic terms developed in many cases from the names the producers gave their products. Very often it is not even known to specialists that, for example, LOM (laminated object manufacturing) is the registered trademark of Helisys[2] and is therefore a brand name and should not be used as a generic term for layer laminate processes (LLM). As long as a process was the only one of its kind on the market, the producers did not really mind if their name was also used as a generic term. However, now that other competitors have entered the market with other processes (e.g., fused deposition modeling, FDM, which is a FLM process, is the registered trademark of Stratasys), most producers observe punctiliously that their brand name is used only for their own products and not as a generic term. Therefore, technological generic terms and producer-specific brand names should be kept painstakingly apart.

The market is changing rapidly. It is to be expected that within a short time further marketable systems will become available that are currently still being tested or developed.

---

[1] The "M" in the abbreviation stands for either "modeling" or "manufacturing" depending on the author.
[2] Helisys went off buisness in spring of 2001 succeeded by Cubic Technologies, see section 3.3.3.2.

Only industrially used systems are to be considered here. These include prototypers that are commercially available (or that will be available, according to the producer, within a short time), which provide a professional service such as the one typical for machine tools and for which updates and upgrades are guaranteed.

3.1 Rapid Prototyping Process Chain

Usually, when the decision for a rapid prototyping process has to be made, the prototyper itself and its suitability for special tasks are scrutinized. Mostly, however, the geometry data are neither available in the form the prototyper requires nor can the model be used in the form in which it leaves the prototyper.

Therefore, the decision for a certain process also determines the entire rapid prototyping process chain. The process chain is outlined in Fig. 3-1. This illustration is based on the principal interrelationships in Fig. 2-2.

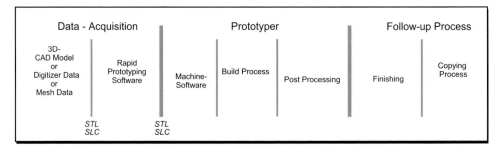

Figure 3-1 Rapid prototyping process chain

The starting point is a complete 3D data set (solid model).

This is produced in most technically relevant cases with the aid of a 3D CAD system. The prototyper can be addressed directly via an STL interface. In practice, users often prefer to add a rapid prototyping software that enables the observation, orientation, repair, and scaling of the models. The aspects of data systems technology are described in Section 3.2. Technical data can be found in Appendix A2-1.

Prototypers are described in detail in Section 3.3. In addition to the prototyper itself, the section concerning prototypers in the process chain also covers the necessary software for machine control and process typical post-processing. In Section 3.3 the prototypers categorized according to the following:

• Type, manufacturer, marketing, short description, application, development state

• History, development partners/strategies

- Data formats/software, principle of layer generation, design/construction, material/build time/accuracy

- Post-processing

- Process typical subsequent processes (finishing and casting techniques).

The common properties of several prototypers of one class are mentioned at the beginning of each chapter (Sections 3.3.1.1., 3.3.2.1., and 3.3.3.1).

Technical data and information concerning manufacturers, prototypers and materials are brought together in the Appendices A2-4 to A2-14.

Prototypers usually supply one prototype (piece no. 1) made from one process-specific material while the user very often expects a large number of prototypes made from materials as similar to series materials as possible. This is usually achieved by duplicating the rapid prototyping original models or by rapid prototyping processes used for the production of tools (see Chapter 4)

The most important casting processes for the production of prototypes in plastic and metal are described in Section 3.4. Material information for casting resins can be found in the Appendix A2-25 "RP-Materials and Casting Resins."

3.2 Data Technique

To address generative production installations the geometric data only need to be transmitted in STL format and the CAD system used should accordingly have an STL interface as output format (see also Fig. 2-2).

The quality of the models depends to a large extent on the quality of the data, that is, whether they are transmitted completely and without errors and how well and efficiently the geometric data are prepared enabling an error-free, build time and quality optimizing process to pass through the prototyper. With a ratio of preparation time to build time of approx. 3:10 [DAH98], an intensive preparation with optimization is always better than an unsuccessful build process.

When considering rapid prototyping installations and processes it is therefore very important to devote also a great deal of time to learning as much as possible about CAD systems, interfaces, and program systems for handling and processing geometric data.

As long as the model data are available in a common convertible format for the application of rapid prototyping processes their source is not important; for the construction of rapid prototyping models it does not matter whether the data record was originally based on a 3D CAD model construction, on digital images, or on measured data (reverse engineering).

Problems with the interface and errors in the data record should not be underrated, but they are not a continuous problem when the cooperation between technical designer and model maker is well coordinated. The data formats and data flows should be tested and standardized within the company as well as in cooperation with external service people. When time schedules are tight, new formats, programs, interfaces, and changing partners are a problem, not only in prototyping.

CAD Systems and CAD Interfaces

Although nearly all CAD systems in use today possess an STL interface it may be of advantage to transmit the data via standardized (neutral) interfaces (and thereby in different formats). When technical designers order prototypes they use neutral interfaces in their daily routines. The STL format as "special interface" is rather unusual. The processing of reversed data is also easier.

From the beginning, interfaces have grown with CAD systems. As a result, a large number of different interfaces exist, today, some of which are competing with one another. In any case, for rapid processing it is an advantage that only geometric data are required and none of the additional information that is usually included in complete CAD data records.

Neutral Interfaces

The most important interfaces are:

- IGES   Initial Graphics Exchange Specification
- VDAIS   Verband der Automobilhersteller (German Association of Car Manufacturers) – IGES interface
- VDAFS   Verband der Automobilhersteller (German Association of Car Manufacturers) – surface interface
- DXF   Drawing Exchange Format
- HPGL   Hewlett Packard Graphics Language
- SET   Standard d'Échange et de Transfer (Standard for Exchange and Transfer)
- STEP   Standard for the Exchange of Product Model Data.

IGES defines a worldwide standard for a geometric interface, but has many variants that must be described exactly.

VDAIS is an IGES interface whose number of different elements was restricted by the Verband der Automobilhersteller (VDA) (German Association of Car Producers). It, too, has a number of variants; it is necessary to establish on a case-for-case basis on which interface formulation the data transmission is based.

VDAFS is specialized in the transmission of free form surfaces and is therefore of outstanding importance, especially in the automobile industry.

HPGL is a contour-orientated plotter format that is also used for rapid prototyping processes with a CAD, because of its ability to generate contours directly.

STEP is a newly developed interface that has been in the test phase for some time and that, in addition to geometric data, also transmits other information. STEP, therefore, represents an approach to the transmission of actual CAD program systems and not merely geometric information (with additions) between CAD program systems.

Rapid Prototyping Interfaces

STL Interface

Data transmission to the rapid prototyping system in most cases passes through an STL interface, the so-called „front end" software for the preparation and control of the model-making process (rapid prototyping software). The STL-format has become an industrial standard today.

Figure 3-2 shows how in STL formulation the actual contour is converged by triangles and how this is altered with the contour.

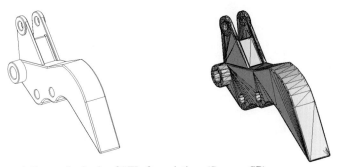

Figure 3-2   Triangulation as the basis of STL-formulation. (Source: CP)

SLC Interface

Alternatives to STL interfaces are contour-orientated interfaces (SLC). Although they are basically 2D data, their importance will continue to grow to the same degree as parametric solid modelers gain acceptance as CAD systems. The fact that SLC formatted layers cannot be scaled later is a disadvantage that will lose its importance only when the slicing is done interactively and as a matter of routine out of the CAD.

The data format selected not only influences accuracy and build speed, but also the process chain, the division of labor, and responsibilities. When transmitting STL data, design and model making are completely decoupled. However, the transmission of SLC data requires that subsequent operations are carried out in the same CAD system. This means that the technical designer must take over specific tasks of the model maker (orientation, cutting, calculation of supports). The responsibilities of technical designers and model makers are no longer clearly defined; they continuously merge into one another (Fig. 3-3).

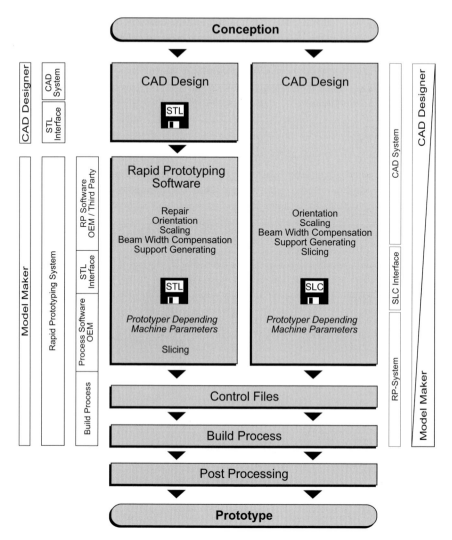

Figure 3-3 "Rapid prototyping" process chain and responsibilities. STL formulation (left) and SLC-formulation "direct slicing" (right)

Rapid Prototyping Software

The simplifying image of the prototyper as a 3D copier is good enough for the theory of product development but it is distinctly unhelpful when considering the process itself. Prototypers are generative production installations with the character of machine tools.

When geometric information is transmitted as a data set, the model maker receives a complete data model of the component. The data records are the basis of communication between the technical designer ("design") and the model maker ("production"). They are also the basis for calculation and preparation of the production process. The model maker must be able to judge the data and he needs therefore to be able to visualize them. In most cases he then also uses a CAD system. The back transfer of data from the CAD via standard interfaces is more effective than the transmission of altered STL data, especially when alterations are necessary which usually require the approval of the technical designer.

The requirements of a modelmaker, however, go far beyond the standard CAD functionality. He must not only visualize the data record and therewith the later model, but he must also be able to orientate it optimally in the build chamber and, if necessary, to scale it. Common errors in the STL data record such as interrupted surfaces need correction ("repairing"). Larger data records are cut into pieces to be built as parts of a model and later joined to a complete model. Supports (Stereolithography, FDM) or bases (for some sintering models) or separating cuts and planes (LLM-Kira) require definition.

For quality management, it is important to be able to take random measurements from the drawing and to check them against the model. This applies especially to mold making (rapid tooling): for the definition of parting planes, for the generation of offset surfaces, for the addition of machining tolerances, for the placing of drafts, for the inversion of positives to negatives, and for the scaling of the models.

Therefore, programs are needed that cover the whole range discussed, from a simple CAD program to special programs for the calculation of supports or the simulation of production times.

CAD systems are in some respects too complex for these tasks but on the other hand their functionality is usually inadequate. All producers of prototypers provide software which is usually adequate for making models. The necessary „front-end" software providing the above mentioned functions was long neglected by the producers of prototypers in respect to functionality and user-friendliness. As a result, highly efficient independent suppliers (third parties) with system-independent "front-end" solutions have established themselves. As these systems by their very nature had to be kept open for covering different rapid prototyping systems, they were the first to provide neutral interfaces (e.g., IGES, VDAFS, or DXF); in this way they support the direct link of solid-modeling CAD programs, which are flooding the market, with prototypers,.

Especially for those users of prototypers who do not possess their own CAD systems or the specific CAD system necessary, in special cases the use of such open rapid prototyping systems, which are in principle CAD systems on the basis of a surface modulator reduced in its functionality, is becoming increasingly attractive. This software, which is relatively easy to learn owing to its reduced functionality, enables geometric data to be verified and processed.

The most important characteristics of software for rapid prototyping systems are listed in Appendices A2-1 and A2-2. Producer-specific software that can be run only on special machines is discussed together with the corresponding prototyper.

In practice software is becoming increasingly producer- or machine-specific the closer the process gets to the construction process. General geometric manipulations are performed in the CAD, while specialized rapid prototyping processings take place in various programs of the "third party" type. Machine-related operations are carried out using the software supplied by the producer of the prototyper. In the course of the process chain, various programs are used that are coordinated with one another. Many producers have started to sell well established "third-party programs" or to offer them under their own brand name,[1] thereby blurring the dividing lines in every respect. As a result the user finds it necessary to build up his own special program system that matches his requirements. The producers usually give support where they can but they do not supply any complete user libraries.

As the support genereration is especially important for all stereolithography processes, being the largest section of the rapid prototyping family, programs for support generations are also listed together with their system properties (see also Appendices A2-1 and A2-2)

In addition there are special programs. Materialize, a company based in Leuven, Belgium, for example, sells a program system with which computer tomographies (CT's), which are generated by a computer scanner or other mostly medical radiographic processes and are available as layer information, can be converted into 3D build data.

3.3 Prototyper

This section looks at industrially available prototypers. They are listed according to their basic functionals as described in Section 3.1 and arranged within this group according to producer. The descriptions deal with single machines or groups of machines of one manufacturer.

Especially those prototypers are listed that are used in industry, for which upgrades and service are guaranteed and that possess the typical up-time of machine tools. As the border lines are blurred owing to the continuous rapid development, some systems are listed that are on the threshold of commercialization. Rapid prototyping is used for more than 15 years, but changes in company reorganization, acquisition or even liquidation happen as in every other buisiness. Therefore even some prototypers are described that are not produced any longer because many systems are still in use.

The description of prototypers follows the structure outlined in Section 3.1. The description of the prototyper itself is given most attention. These are unbiased reports on the specialties of each machine. The order of the reports and their length do not pass any judgment.

All aspects of the technical realization of the principles described in Chapter 2 that are common to all prototypers are discussed beforehand.

Stereolithography is deliberately discussed in greater detail than other contributions because, being the first and still the most accurate commercialized process, it is the

---

[1] It should be taken into account that the programs are not always full versions and sometimes necessary additional modules need to be bought.

benchmark for many others. The more important reason, however, is that the exact description of stereolithographic model making is suited to engendering a feeling for generative processes which, in many aspects, can be applied to other processes.

The reader is well advised to read this passage even if he is not interested in stereolithography.

3.3.1   Photopolymerization – Stereolithography (SL)

3.3.1.1 Machine Specific Basic Principles

The industrial application of the solidification of liquid monomers by photopolymerization is known as stereolithography. 3D Systems calls the process "StereoLithography;" EOS called it "Stereographie." Both terms are registered trade marks of the respective company.

Stereolithography is the origin of all industrially used rapid prototyping processes and, with more than 2500 installations (at the end of 2001) worldwide (more than 600 in Japan), it is one of the most widely used industrial application.

The oldest and most widely used system is based on the principle of point-by-point solidification by means of a laser scanner exposure. The lamp-mask process (solid ground curing) is described in Section 3.3.1.5. Nowerdays new processes based on photo polimerization are published: EnvisionTec uses a DLP-projector to cure the model upside down through a glass plate (Section 3.3.1.7.) and Objet takes printheads to apply the liquid monumer (PolyJet process) which is continuously hardened by a lamp (Section 3.3.1.8.). 3D-Systems (originally Optoform's (Nancy, F) paste polymerization process) applies a pasty monomeric resin using a docter-blade. Then each layer is shaped and polymerized by a laser. The process is not discribed in detail because it is not commercializd until now. It is obvious that the last two processes on one hand can be named "stereolithography" but on the other hand "3D-Printing-" or "Extrusion-Processes" as well.

In theory, liquid resin stereolithography machines consist of a container that is at the same time build chamber and storage space, filled with a liquid monomer. It is equipped with a build platform moveable in the z-direction and a laser scanner unit, which "writes" or projects the current layer information on to the surface of the resin bath thereby curing one layer. The platform carries the model (on supports that hold projections and unconnected model parts in place and facilitates the defined lowering of the model on to the build platform and its removal from the platform) and lowers it by one layer thickness after a layer is solidified. A new layer is then prepared (recoating), again exposed and thereby solidified, and so forth. In this way the model "grows" layer by layer from bottom to top (Fig. 3-4).

All laser-supported stereolithography processes aim to solidify one layer resulting from single solidifications, so-called voxels. Voxels, which ideally have the shape of a paraboloid of revolution, result from the energy distribution in the laser beam and the penetration characteristics of the resin. To achieve the necessary compound stability, the voxels of one layer and those of two neighboring layers penetrate one another (overcure). The actual penetration depth (called cure depth) of the laser is thereby deeper than the layer thickness (see Fig. 3-5) and the generation of one layer and its interlocking with the lower preceding layer takes place simultaneously.

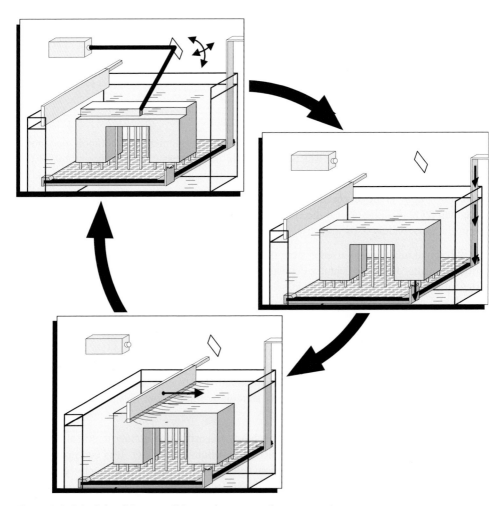

Figure 3-4 Principle of the stereolithography process (laser scanner)

In practice, the interaction of laser performance, beam parameter, scanning speed, and material characteristics (type of resin) decides whether a voxel structure really grows in the layer, or whether quasicontinuous paths are written.

For a rapid and exact polymerization different producers use different exposure strategies. Basically the layers are first contoured by borders and then internally cured by suitable hatches. The beam diameter is compensated for an exact generation of borders. To achieve this the laser path is diverted by half its beam diameter off the correct contour into the component (beam width compensation or line width compensation).

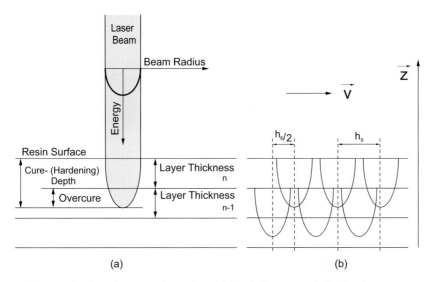

Figure 3-5 Interaction laser beam - resin surface. (a) Single beam voxel. (b) Voxel structure

The volume of the resin decreases due to polymerization and the model shrinks. The problem of shrinkage has been considerably reduced since the change from acrylates to epoxy resins (linear shrinkage: acrylate = 0.6% / epoxy = 0.06%). However, epoxy resins need up to three times higher exposure energy. There are several build strategies that are used in addition to the optimization of process parameters to counteract the effects of shrinkage. One possibility is not to join opposite walls continuously, but to generate gaps periodically (as shown in Fig. 3-6a) which prevent deformation due to their internal stress (retracted, 3D Systems).

Figure 3-6b shows the situation in a real component. The cure tracks do not touch one another (hatch distance, 0.26mm) and leave sufficient distance to the border (retracted, 0.2mm). The pattern was made with the aid of STAR-Weave technique (3D Systems).

The parts generated by laser scanner processes are of a relatively low stability during the build process (green strength), similar to that of gelatine. Overhangs, such as projecting elements or cantilever walls therefore need support by support constructions if they exceed a certain size or angle (Fig. 3-7).

The function of supports as substructure for the whole model is outlined in Fig. 3-7a. Support constructions such as these coordinate the plane of the build platform with that of the recoaters. They also enable the model to be removed from the build platform after the curing process is finished.

It is the main function of these structures to support projecting parts of the construction and also to pull other parts down which due to shrinkage tend to "curl" upwards. The supports shown in Fig. 3-7b are gussets, which are used especially for supporting rectangular geometric ramifications. Figure 3-7c shows a construction known as an "island." This aids the positioning and supporting of elements that start in later layers and join the model even later. It would be impossible to build handles of cups, for example, without such supports because they would start "in the air."

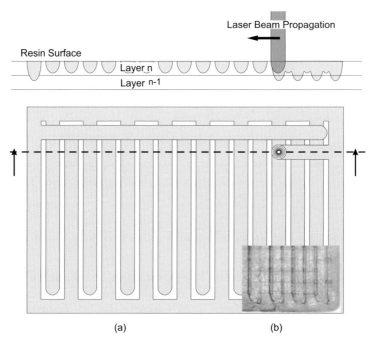

Figure 3-6 Build style to reduce internal stress (retracted). (a) Principle (b) Real structures of 3D System's STAR-Weave build style. (Source: CP)

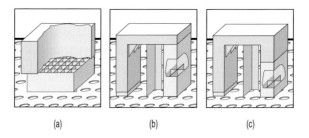

Figure 3-7 Support structures. (a) Substructure. (b) Support. (c) "Island"

These supports are (automatically) generated during data processing. That causes the amount of data to grow tremendously, especially with STL formulations. It has to be made clear that each support is materially a volume element and must be made not of just two but of twelve triangles as in a cuboid (Fig. 3-8a). Depending on the model, effort expended in generating and removing the supports (post-processing) can be enormous. In addition, the rather solid support structure can damage the model while it is removed using tools such as knifes.

Therefore a new support strategie called "fine point" was developed by 3D Systems and registered as the company's trademark. Figure 3-8b shows a model of a plasma welding device after the SL built process but still on the plattform and with fine point supports.

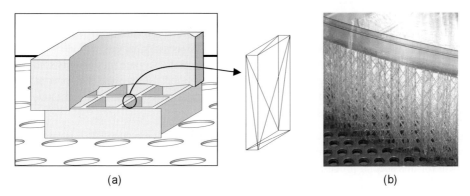

(a)   (b)

Figure 3-8 Supports. (a) Classical support construction as volume element. (b) "Fine point" support
structure (Source: 3D Systems)

For the stereolithography processes of all manufacturers, build strategies have been developed with a higher proportion of hollow space volume to increase the build speed and to decrease the amount of material. These strategies are especially aimed at being able to employ models directly as lost models in a precision casting process (burning from ceramic precision casting molds). Walls that are in reality solid are designed as hollow walls connected by rod-type bridging elements. In addition, so-called skins need to be introduced that close the model at the top and the bottom (in the z-direction).

Hollow stereolithography models require openings through which the monomer that has not been cured can be drained. These openings must be closed again to ensure that no ceramic slurry can flow into the model and cause casting faults, if used as a precision casting master.

When the last (top) layer has solidified, the build process is finished. The model which is now completely dipped in the monomer is removed from the vat, enabling superfluous resin to run off and drip back into the resin container. If the machine is to be used economically it is better to place the model into a separate, preferably heated container to drain off. There are variants in which the resin is drained off by a process similar to the centrifuging of honeycombs. This is especially beneficial for hollow walls.

The procedure known as *post-processing,* in which the models are cleaned, the supports removed, and the models post-cured, is the same for all laser-supported stereolithography processes. The procedure is shown schematically in Fig. 3-9.

All models are first cleaned (acrylates with TPM); after cleaning, the solvents are removed as far as possible (TPM with isopropanol).

In laser-supported stereolithography processes the component is usually polymerized by up to about 96% (depending on the penetration depth). The models, therefore, are not post-cured in the prototypers, but in special UV ovens (post-curing ovens). In the lamp-mask process as well as in the resin printing or paste polymerization process the models leave the prototyper completely cured.

The supports are removed before or after the curing, depending on the complexity of the model and the demands on accuracy and material. The supports are removed manually, a process that demands great care and that is therefore time consuming. It is thus advisable not to generate more supports (automatically) than is necessary, as this will slow down the build process of the model. To achieve higher accuracy with acrylate models it is sometimes advisable to cure them completely with the supports and remove the supports after the curing is completed. Thanks to the fine point support structure, even filigreed model parts can be cleaned whihout the risk of damage.

Another post-processing procedure is the implementation of a mechanical finishing by sand blasting, polishing, milling, and so forth as far as is necessary for achieving the qualities typical for this special process.

A differentiation is made between process-orientated post-processing and *finishing*, which means process-independent further treatment of the surface. This could, for example, be done with varnish, fillers, and stoppers of various kinds but also by electroplating or plasma coating and is not a rapid prototyping specific process step.

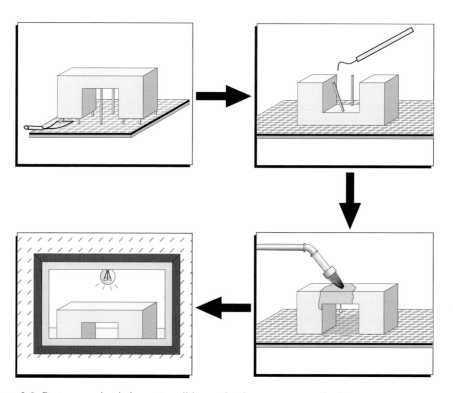

Figure 3-9 Post-processing in laser stereolithography (laser scanner method)

The producers of prototypers usually do not offer (but sometimes recommend) equipment and tools for finishing (polishing, sand-blasting, varnishing, etc.) and for subsequent casting processes. These aids must be chosen by the interested user himself.

Stereolithography models show the highest degree of details and the best surfaces. The modest mechanical-technological properties of resins are a disadvantage. Optimal polymerization comes first and foremost while all other properties such as resistance to tension, temperature stability, and so forth are of secondary importance.

A promising strategy is to separate the properties: The stereolithography process enables a reproduction in great detail. The required mechanical-technical properties are achieved afterwards by molding processes. For the production of plastic prototypes, therefore, the process chain – stereolithography – vacuum casting – is well established as a standard if the number of pieces required lies between 5 and 15 (sometimes more) and the simulation of the later injection molded material by castable or pourable (PU, polyurethane) plastic material is acceptable (see also Section 3.4).

Metal models are produced by making wax models by vacuum casting, which are then melted out by classical precision casting processes. As an alternative, hollow stereolithography models can be used directly.

Tools are produced by casting processes or counter-casting processes with metal-filled resin or by low-temperature coatings (see also Chapter 4).

3.3.1.2 StereoLithography Apparatus (SLA) – 3D Systems

Viper si2 (SLA250)/ SLA3500/ SLA5000/ SLA7000

3D Systems Inc., Valencia, California, USA

Short Description: The stereolithography machines produced by 3D Systems work with the laser scanner process. The model is generated on a platform. After contouring and haching the upper layer is completed and the platform is lowered into the resin bath by one layer thickness, enabling a new resin layer to be applied onto it's surface. The process is repeated until the model is finished.

The removal of the supports, the further post-curing in an UV oven, and the finishing of the surface are process steps that take place outside the prototyper and that are essential for producing high-quality models.

Range of Application: Geometric prototypes, functional prototypes (technical prototypes).

Tooling directly via ACES injection molding, indirectly via casting processes (Keltool).[1]

Development State: Commercialized.

---

[1] Investigation on and further development of the 3D-Keltool process was stopped in November 2001 by announcement of 3D-Systems. Existing customers will be supported futher on and new ones will be served based on the technical status quo. Details on the 3D-Keltool process see Section 4.2.2.2.

History: In 1986, 3D Systems was established by Chuck Hull and others, who was granted the patent of the "Stereolithography Apparatus" in 1984 and who is considered to be the father of stereolithography. In the mid-1980s the first stereolithography machine (SLA1) was built and, based on this concept, the first commercial stereolithography machine (SLA250) was introduced in 1988, which started the tremendous development of rapid prototyping processes. The first stereolithography machines were shipped abroad, especially to Europe and Germany around 1990.

The SLA250 is considered the forefather of all stereolithography machines, and in fact of all prototypers. Its basic principle for construction can still be found in all laser-supported stereolithography and laser sintering machines as well as in most other prototypers. Even the newest development, the SLA Viper si2 (2001) basically is still an SLA250. Together with the larger versions, the SLA3500, the SLA5000, and the SLA7000, it belongs to the same family of stereolithography machines. For technical data please refer to Appendices A2-4 – A2-7. Figure 3-10 shows the SLA250. The smaller version SLA190, the SLA350 and the SLA500, the predecessors of the SLA3500 and the SLA5000, and the SLA250 are no longer being built.

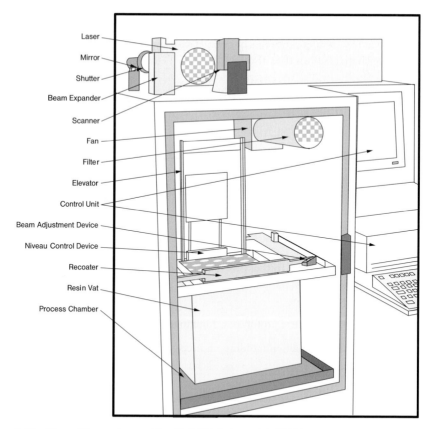

Laser
Mirror
Shutter
Beam Expander
Scanner
Fan
Filter
Elevator
Control Unit
Beam Adjustment Device
Niveau Control Device
Recoater
Resin Vat
Process Chamber

Figure 3-10a Stereolithography machine SLA250 (schematic), 3D Systems

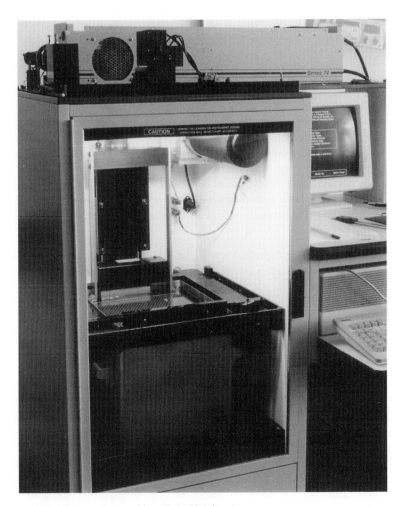

Figure 3-10b Stereolithography machine SLA250 (photo)

Development Partners/Strategies: The aim of development is to create a family of powerful prototypers that, in connection with suitable casting processes and process chains, can produce functional prototypes, technical prototypes, and series components for industrially relevant parts with characteristic dimensions ranging typically from several centimeters up to about 0.5 m. The Viper si2 and 250 series are the prototypers for beginners and intended for occasional use, whereas the 1000 series was optimized for productivity in the sense of maximal build capacity. Development partnerships exist between rapid prototyping software providers (Solid Concepts, Materialize, and others). A promising partnership with VANTICO, a supplier of resins was seezed in 2001. Besides the RP-cure resins (RPC, Switzerland) 3D Systems now run their own ACCURA materials line.

Data Formats/Software: The geometric data of the prototype are usually available in the so-called STL format. In this format the surface of the solid is converged by triangles.

3D Systems offers a very extensive front-end software called LIGHTYEAR which replaces the well know MAESTRO. It includes a number of subprograms which are no longer given specific names but in the following paragraphs they are described with their functionalities authentic for the SLA250.

By means of the *View* program the STL solid can be viewed and turned, scaled, mirrored, and shifted. This is a very important aid because the data record in STL representation (triangle surfaces) can never be judged accurately enough. In addition, it is important to be able to place components independently of the techincal designer's CAD system in the best possible way for stereolithographic production. The View program allows additional repairs and constructive additions to be made so that for small alterations it is not always necessary to return to the CAD system.

Furthermore, the latest version (Solid View, Solid Concepts) facilitates the registration of characteristic measurements and, when the measurements of the finished model are taken later, a quality check is possible.

The program *Bridgeworks* (Solid Concepts) enables supports to be generated automatically. The automatic support generator is especially helpful for standard tasks. For complicated components, to reduce the construction requirements and thereby saving time and reducing the necessary finishing processes, the experienced user should use the possibility of manual interaction.

The data records of supports and prototypes are loaded separately into the parts manager *Partman* where they are separately processed. The recoating parameters as well as the component parameters are fed into Partman. The recoating parameters include the number and speed of sweeps, the speed of the elevator, the rate of acceleration of the elevator, the elapsed time between the end of the scan and the start of the elevator movement, and the elapsed time between the end of the elevator movement and the start of the scanning. The component parameters include layer thickness, border overlappings, hatch overlappings, hatch distances, cure depth, space between fills, direction of fills, distance from border, beam compensation, build style. The component parameters are defined separately for each set of supports. Component parameters that are already recorded can be read and edited.

After the STL file has been processed in Partman, a slice file is generated from it. In this data record, the geometric of the solid is stored in layers. Afterwards another four data records are generated automatically that contain the control data for the SLA. The vector file contains all contour data for controlling the laser. In the recoat file the data concerning the cure depth are recorded from which the overlapping of single surfaces is derived. The parameter file contains general information such as the shrinkage factor of the resin or the position of the component on the build platform. The layer file contains information such as the layer thickness and the hatch distances for single supports and components. To enable the start of the production process these four data records are complemented by the resin parameters (C_d, D_p).

For the SLA3500 and 5000 only two files containing the contouring information are prepared. The processing of border and hatch structures takes place parallel to the build process in the machine. The machine-oriented part of the software is called *Builtstation*.

The SLA7000 probably uses basically the same software [Builtstation 5.0 (2000) is included in the Lightyear package]. It is much faster than the MAESTRO software with regard to 3D reproduction as well as data checking. To control the adjustment of the beam in the SLA7000, fundamental detailed manipulations are necessary that are not noticeable to the user.

The Principle of Layer Generation: The procedure for prototype production is discussed in detail in the following paragraphs based on the SLA250. This procedure is, with reservations, similar in all stereolithography processes and in its essential features to many rapid prototyping processes based on the laser scanner principle.

To establish a firm connection between the supports and the elevator platform the first layer is cured into the perforated pattern of the platform. Controlled by the data of the vector files, the laser beam cures the model together with the supports in layers. The speed of the laser beam varies between 0.762 m/s (SLA250) and 9.5 m/s (SLA7000) depending on the cvure depth. When a new layer is started, the laser runs along the borders of the component (borders). Then the plane between the borders is hatched (hatches). The distances between the hatches are so narrow that only a very small portion of the resin stays liquid. In the 7000 series the curing is sped up by running only over the contours very finely. For filling the contours the laser beam diameter is widened significantly (Dual-Spot).

To connect the single layers with one another the cure depth of borders and hatches is made bigger than the layer thickness which results in an overlapping (overcure) of about 0.03 mm; as a result of this, downward facing surfaces that start on an arbitrary z-coordinate during the build processes become larger in the z-direction. The machines therefore are fitted with a z-correction to compensate for this effect.

Figure 3-11 shows a build platform with examples of models from the applications in Section 5.1.4.

Those parts of the layer representing the surface (skin) of the component can be filled additionally to the hatching. The filling parameters enable the component to be cured more efficiently than in the case with the hatching alone, giving the component surface a higher stability. Figure 3-12 points out the differences between border, hatch and fill.

The minimum layer thickness is 0.15 mm (SLA250/30A), 0.1 mm (SLA250/50 and SLA250/50HR), 0.05 mm (Viper si2, SLA3500 and SLA5000), and 0.025 mm (SLA7000). According to different build styles the layer thickness can vary.

After a layer is cured, the build platform with the model is lowered by one layer thickness and a new layer of resin is applied. This procedure is called recoating. At first, 3D Systems used predominantly the method of passive recoating by means of a blade type recoater. It's main function is to smoothen the resin surface. Modern machines, with the exception of the SLA250/30A, are fitted with an active recoater that is specifically designed to apply material.

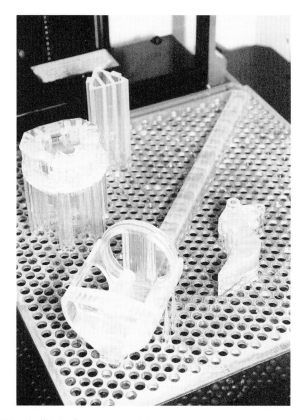

Figure 3-11   SLA250 – build platform with models

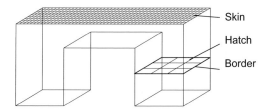

Figure 3-12   Definition of border, hatch, and fill (or skin)

With a passive recoater, following a 3D patent (deep dip), the component is lowered deeper than the thickness of one layer under the resin surface and is moved back a few seconds later toward the direction of the next layer but a bit further up. Owing to the viscosity of the resin, a small resin mound is formed on the component. A wiper blade then smoothes the resin surface on the component after which the build platform moves back to its exact position. By this method less flow is generated in the resin than by wiping directly over the entire surface of the resin.

So-called "trapped volumes" present a problem. These are volumes opening upwards only and, owing to the model contour, they are isolated from the rest of the bath. Such trapped volumes sometimes require highly differing recoating parameters that are difficult to assess. It can easily happen that the borders of such trapped volumes press the "bow wave" of the wiper underneath the wiper blade, causing an uneven resin surface behind the wiper. As the wiping procedure for that layer is then finished the laser will cure the wavy structures. Apart from the resulting construction fault, there is also the danger that during the next recoating process the wiper will run into the "mound" remove layers, and thereby damage the model. It would help to avoid this problem if the part were placed in a different way, or if drainage openings securing the fluidic circuit with the rest of the bath were installed.

Active recoaters are not only an elegant remedy for "trapped volumes," but they are also faster and cause an even surface by applying material. Therefore all modern prototypers are equipped with the active Zephyr Recoating System. It was patented by 3D Systems and shows a hollow wiper with an integrated resin reservoir, open towards the bath. The surface of the reservoir is slightly underpressurized, keeping it filled with resin from the bath surface. The pressure ratio is balanced in such a way that, when wiped over solid surfaces, resin sticks to these structures and is thus applied while from the open surface resin is sucked into the reservoir. This system is not only more exact than the passive one, but it is also much faster because the "deep dip" is eliminated and because the bath usually needs to be crossed only once. In a so-called "smart sweep" process the wiping is limited to the y-dimension of the model cross section which gives the process additional speed. In contrast to systems that transport the resin by pumps, the Zephyr system is relatively easy to clean if the resin is changed.

The build style originally developed for acrylates, *STAR-Weave*, was modified as epoxy resins became more commonly used. The build style known as accurate clear epoxy solids (ACES), enabled 100% cured epoxy solids to be made. To achieve this, hatches especially are put together so tightly that they partially overlap.

Solid models are needed especially when a higher (pressure) stability is demanded, for example, in rapid tooling applications (AIM, ACES injection molding), or when optical properties are especially important.

Quick Cast is a special build style that combines both two important requirements for the construction of hollow models: good drainage and the highest possible hollow volume quota. This build style enables to make models with more than 80% void volume (in relation to a solid model) which can be preferably used for precision casting applications. Specially suited resins are used that generate smooth walls and drain off easily because of their relatively low viscosity. (Accura and RP Cure Resins wih various properties).

Figure 3-13 shows how the hatching builds up on a basic algorithm of 0°/60°/120° in a horizontal x-direction. In a second hatching step the pattern (Fig. 13a) is turned in relation to the first step (Fig. 3-13b) and repeated, shifted in the y-direction. A honeycomb pattern results (Fig. 3-13c) with connecting hollow spaces in the z-direction that allow a simple drainage of the model and the complete evacuation of remaining liquid monomers.

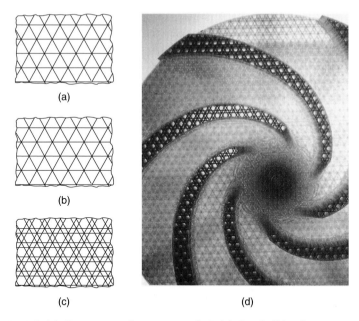

(a)

(b)

(c)   (d)

Figure 3-13 (a–c) Quick Cast construction strategy. (d) Quick Cast build style on a pump impeller.
(Source: 3D Systems)

System Type/ Construction: In all systems the entire process chamber, in which the actual model is generated, forms a single structural unit together with the process and control computer and the laser scanner device. Figure 3-10 shows an SLA250 with the upper casing removed so that the laser as additionally procured independent component, the beam guidance, and the scanner unit are clearly visible. The assembly of machines from as many common components as possible and resulting in an extremely low manufacturing depth is another hallmark of most producers of rapid prototyping systems. The relatively weak HeCd lasers of the 250 systems with 12 to 60 mW are replaced in larger systems by the Nd:YV04 solid-state lasers Nd:YV04 solid-state lasers that can bring up to 800 mW onto the surface of the vat. The Viper si2 also uses a ND:YV04 laser with 100mW output power. The elevator, which together with the perforated build platform forms the basis for the model-making process, dips into the resin bath from above and can be withdrawn completely. In this way the resin container can be changed in all state-of-the-art machines.

The SLA7000 (Fig. 3-14) is fitted with an 800-mW laser, which has an adjusting mechanism enabling the user to select one of two different beam diameters. This innovation shortens the build time which is limited by the repetition rate of the laser. The build time is prolonged by the time-consuming hatching process but it is effectively shortened without the scanning speed being raised when the beam diameter is widened. It is not possible to raise the maximum scanning speed in excess of the product of repetition rate and spot diameter if a coherent model is to be produced. As long as the repetition rate cannot be raised significantly above 10 to 20 kHz, the only alternative is to widen the spot diameter. Current technical data allow us to assume a repetition rate of about 10 kHz.

Figure 3-14 Stereolithography apparatus SLA7000. (Source: 3D Systems)

Materials/Build Time/Accuracy: The quality of the model depends not only on the type of resin but primarily on the interaction of the laser and the resin. For this reason an interchangeable container alone does not guarantee an all-purpose machine. The less powerful helium-cadmium lasers, standard in smaller machines, require extremely long build times for epoxy resins. Conversely, machines equipped with high-powered lasers working with acrylates tend to cause filamentations that are not only messy but also contaminate the resin and lengthen the cleaning procedure. Even among epoxy resins there are considerable differences.

3D's family of SL prototypers uses all in all three types of lasers (the older 350 and 500 systems have argon-ion lasers) each with different performances. There is a large range of resins, therefore, each one of which is optimized for special model requirements and also for particular lasers and their performances. The disadvantage is that different machines running at the same time for productivity reasons cannot be fed with the same resin. The result of the build process depends strongly on calibration as in all rapid prototyping processes. The maintenance personnel, therefore, use a calibration plate with which they trace the position of the laser beam by means of diodes and analyze the data on-line. The process is very effective.

In the SLA3500 the resin can be refilled and changed easily by special cartridges.

The typical construction showing a combined build chamber and resin reservoir has, especially for big stereolithography machines, the disadvantage that there is a large amount of resin to be kept in the machine (in an SLA5000/7000 this can be worth as much as $ 40,000). Because resins tend to age and require an accurate climatization, the machines must be used continually.

Post-Processing: For all 3D stereolithography prototypers the post-processing is identical to the stereolithography-typical post-processing described in Section 3.3.1.1.

Process-Typical Follow-Up: All stereolithography-typical finishing techniques and casting processes are possible (see Section 3.3.1.1).

Precision casting parts are obtained from stereolithography masters made by quick cast build style. This implements hollow structures and prevents the ceramic shell from cracking while it is fired. When using these hollow structures in a precision casting process they need to be drained and closed completely afterwards. Drainage openings must be provided or mechanically drilled to allow this. Before the precision casting starts, these openings must be closed again. Figure 3-15 shows a cylinder block beeing build using the Quick Cast build style and the corresponding casted part (Daimler-Chrysler AG). Owing to the optimization of the curing parameters it is necessary to be mindful that the relatively thin skins do not "collapse."

Figure 3-15 Cylinder block beeing build using the Quick Cast build style and the corresponding casted par. (Source: 3D Systems / Daimler-Chrysler)

Molds and tools are produced directly by AIM (ACES injection molding) or indirectly using 3D-Keltool [DIL84] (Section 4.2.2.2).

3.3.1.3 STEREOS – EOS

STEREOS DESKTOP, STEREOS/300/400/600/MAX 600

Electro Optical Systems GmbH (EOS), Krailling-Munich, Germany

The EOS Stereography machines are not longer produced. Because they show a still quite competitive performance and because there are still many machines in the market they are mentioned here.

Short Description: The STEREOS stereolithography machines made by EOS use the laser scanner process. The model is built on a platform that is lowered into a resin bath. After the current topmost layer has been contoured and crosslinked a new resin layer can be applied to the surface of the model. The procedure is repeated until the model is finished.

Outside the prototyper the model is finished by removing the supports, post-curing in the UV oven, and finishing the surface. These process steps are essential for the creation of a high-quality model.

Range of Application: Geometrical prototypes, functional prototypes (technical prototypes).

Castings and tools are produced indirectly by means of casting processes.

Development State: Commercialized. Production and sales stopped in 1997.

History: EOS was established in 1989 and introduced its first laser stereolithography machine (called stereography machine), the STEREOS 400, in 1990. By 1997 the available number of machines had been increased tremendously and included the whole range of stereolithography machines from a quasi office-suitable desktop machine up to the STEREOS MAX 600 with an Nd:YAG laser and active recoater.

In 1997, EOS stopped the production and sale of stereolithography machines. There was a patent exchange with 3D Systems. EOS took over all the patents from 3D Systems concerning laser sintering while 3D Systems received all the patents relating to stereolithography from EOS. EOS sold the stereolithography line to 3D Systems and withdrew completely from the stereolithography market. Since then the company has concentrated on the development and production of sintering installations.

At the time when the marketing was discontinued, a number of stereolithography installations of EOS, some of which were very new, were in operation and 3D Systems promised to take over the maintenance of these machines. They are therefore still described here.

Until it withdrew from the stereolithography technology, EOS was the only producer in the world that developed and produced stereolithography machines and sintering machines.

Development Partners/Strategies: STEREOS machines are designed as a family, although they are not all based on the same construction concept. Technical details such as the active recoater and the solid-state laser set new standards when they were introduced to the market. The mechanical design fulfills the requirements of a machine tool. The development of the first machine was supported by the German car manufacturer BMW, Munich.

Data Formats/Software: The machine takes over complete and error-free STL data.

The user software supplied by the producer includes a program system that facilitates the most important procedures for the preparation of the build. The EOS software includes the following modules:

- The placer enables the component to be placed optimally in the build chamber by translational and rotational movements. The placer also permits the scaling of components.

- The automatic generation of support structures is carried out in the supporter (see below).

- In the slicer, single layers to be generated later are produced, internal and external structures defined and issued as slice files or SLI files. Sliview™ allows the data to be checked on the monitor.

- Finally, the separator allows extensive STL data records to be checked.

Principle of Layer Generation: The generation of layers is based on the laser scanner principle of stereolithography (Section 3.3.1.1).

One unusual feature is the special kind of support structures generated. Most of the support generators supplied by producers and those from independent suppliers firmly join the supports with the model. Sometimes the contact points are deliberately weakened to facilitate an easy removal of these supports. In any case, they must be removed mechanically after the build process is finished and the contact points usually require proper finishing. As an alternative, EOS has presented an automatic support generator called skin and core supports. The idea is to put a thin skin on a common, quickly build and stable but not too detailed support structure along the joint border between the support structure and the model. This thin skin marks later parting surfaces. The selection of suitable exposure parameters ensures that this skin is cured only so far that it gains sufficient stability to support the model but, owing to a higher amount of monomer, it becomes easy to separate the model manually from the support in these places. This procedure has advantages especially where supports meet complicated 3D model borders, for example, in medical models. Owing to the common separating wall, the fitting of the supports is largely unaffected by model characteristics. Whereas many automatic support generators, especially for complex geometries, require manual operations, this is not necessary with the skin and core algorithm. It should be taken into account, however, that, in principle, the parting surfaces are not able to absorb high loads, that means they can absorb only small tension forces that counter the curl effect.

A further specialty is that the machine is enabled to work with various layer thicknesses and various build styles in one model. In this way it is possible to realize in one model relatively large spatial mesh structures in the supports (core), fine surfaces in the skin, and even hollow spaces, if required, as shown in Fig. 3-16. This facilitates a fast and low-cost construction of the basic geometry without too many details and, at the same time, finely detailed, completely polymerized parts for the representation of important model details (teeth in Fig. 3-16).

For precision casting processes a special build style is available for hollow structures which was developed from the skin and core algorithm .

System Type/Construction: While it was not possible to patent the basic principles of stereolithography because of early publications by Kodama (Japan 1981, see [JTE97]), a large number of details have been protected by patent over the last few years. The recoating system above all is one of them. EOS developed an active coating system as an alternative to the (already patented) passive wiper system. A computer-controlled pump ensures that the correct amount of resin is available at any given time. As the system was introduced as a so-called bidirectional system the coating takes place during one passage over the resin surface. The return trip of the recoating system and the resulting prolongation of the recoating time are thereby avoided.

The active recoating system is directly connected with the software so that the software can respond to special coating situations (trapped volumes).

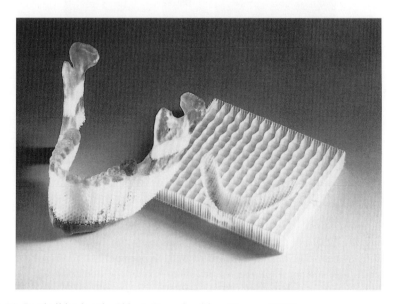

Figure 3-16 Jaw build using the Skin & Core algorithm. (Source: EOS)

STEREOS DESKTOP. At the bottom end of the product range is a machine called STEREOS DESKTOP, which is an interesting alternative especially for beginners in stereo-lithography technology and for users who prefer smaller models. It possesses an active recoating system and all the general characteristics of stereolithography machines as previ-ously mentioned. The advantage of this machine is that its features have consistently been reduced to the absolute necessities. Therefore, it is simple and easy to use and also lower priced.

STEREOS MAX 600. Development targets are a high construction speed and outstanding accuracy. The machine design provides above all a stable base frame resembling that of machine tools, which is the mechanical prerequisite for the targeted model accuracy. In addition, it has an attractive exterior (the judgment of which depends, of course, on indi-vidual taste). The machine is also well accessible from three sides which facilitates an easy exchange of resin containers. Special attention has been given to the ease with which resin containers of various sizes can be exchanged quickly and safely.

Figure 3-17 shows the STEREOS MAX 600.

Figure 3-17 Stereolithography prototyper STEREOS MAX 600. (Source: EOS)

Apart from various details that are rated differently by different users the greatest differences between the EOS and the competing machines lie in the number of usable resins and
in the recoating system. The machine is offered with an optional Nd:YAG (solid state) laser.
The solid-state laser claims a number of advantages over the competing argon-ion (gas)
lasers. It can be assumed that its operative life span has increased significantly by at least a
factor of three. Whereas this is largely compensated for by the purchase price (more than
three times the price of the competing product) its efficiency is distinctly better owing to the
greatly reduced energy consumption which is quickly reflected in the calculation of profitability. Although the solid-state laser needs fitting with a complicated frequency-tripled unit
which impairs its good efficiency, it is still more favorable by a factor of 100 than that of the
competitor, the argon-ion laser. In addition it does not continually lose power during its
running time as does the argon-ion laser. It does not above all have the most awkward characteristic of altering the mode structure during changes of temperature. Disadvantages still
are its relatively low repetition rate, the low efficiency due to the frequency tripling, and its
still relatively high purchase price owing to the small number of units produced.

Material/Build Time/Accuracy: A wide range of materials listed in the Appendices A2-15
to A2-25 can be used in these machines. EOS recommendeds using SOMOS-material
(currently DSM-SOMOS).

Post-Processing: In these installations the post-processing is identical to the stereolithography-typical post-processing described in Section 3.3.1.1.

Process-Typical Follow-Up: All stereolithography-typical finishing techniques and casting
processes are possible (see Section 3.3.1.1).

3.3.1.4 Stereolithography – Fockele & Schwarze (F&S)

Laser Model System (LMS)

Fockele & Schwarze (F&S) Stereolithographietechnik GmbH, Bielefeld, Germany

Short Description: The stereolithography machines from F&S use the laser scanner process. The model is built on a platform that, after the current topmost layer is contoured and crosslinked, is lowered into a resin bath one layer deep so that a new resin layer can be applied to the surface. The procedure is repeated until the model is finished.

The model is finished outside the prototyper by removing the supports, post-curing in a UV oven, and finishing the surface. These process steps are essential for the production of a high-quality model.

Range of Application: Geometrical prototypes, functional prototypes (technical prototypes).

Casts and tools are produced indirectly by means of casting processes.

Development State: Commercialized.

History: Fockele & Schwarze, a spin-off from the University of Paderborn, was established in 1991. It developed its own stereolithography system and introduced it into the market in 1994.

Sales are rather low with a total of approx. 10 machines sold (1998). In the field of stereolithography F&S today acts mainly as a service bureau.

Development Partners/Strategies:. The development is aimed at producing a machine that is as open as possible and therefore interesting for developers of machines, controls, and materials. Users will, of course, also be found at universities and colleges. This applies to the publication of the SCL formulation of the machine as well as for the hardware and software so that the F&S machine is usable as a development platform.

The user may or may not take advantage of the possibility of tampering with the system. The F&S laser model system is a stereolithography machine for everyday use and produces competitive models.

The company is now concentrating on the further development of a metal sintering machine. The machine basically uses the SLPR process and was developed as a result of a partnership with the Fraunhofer Institut für Lasertechnik *(*Fraunhofer Institute of Laser Technology, Aachen, Germany*)*. F&S introduced it in the market in 2002. It will be distributed in context with MCP-HEK (see Section 3.3.2.4.)

Data Formats/Software: The machine accepts HPGL files from the CAD and converts them to contour-oriented build files. Further interfaces are supported.

The system software supplied by the producer is a PC-based software whose program elements can be operated from a joint surface:

LMS Version 3.0 is the control software that takes over the entire machine and process control.

HPGL Convert translates the contour data from the CAD stores it as HPGL-1 slice files into a layer data format developed by F&S.

F&S Editor allows the examination, positioning, and editing of slice files and F&S data.

The *HPGL Editor* allows one to view, edit, zoom, and measure slice data in HPGL-1 format. In addition CAD and slice software is available that can be operated from the CAD platform (workstation).

CSUP is an automatic support generator for contour data that generates support automatically for solid and hollow models (Materialize B.V., Leuven, Belgium).

CSLI generates slice data from STL files and converts them into the F&S data format.

Principle of Layer Generation: The special feature of the F&S machine is its close connection with the software and the direct slicing system in the corresponding CAD system (ProE) [FOC94]. In this way the mathematically exact CAD models are cut and the usual triangulation errors of the STL formulation are avoided. The advantages and disadvantages of this method are discussed in Section 2.2.2.

The generation of the layers is carried out in the way typical for stereolithography machines as described in Section 3.3.1.1.

System Type/Construction: The system in its general design is based on all classical elements of stereolithography machines. Differences exist primarily in the type of layer generation by z-adjustments of the bath. The build platform does not dip into the resin bath from above but is operated as a hydraulic lifting platform albeit with a spindle mechanism from beneath. The displaced volume is mechanically balanced by a counteracting volume of equal size. The coating system is also unusual. It belongs to the family of active recoating systems. The resin bath measuring 400 · 400 · 300 mm has the capacity of bigger machines from other producers. F&S offers the machine with a solid-state laser (300 mW, frequency tripled), but still fits it with a cw-argon-ion laser (up to 100 mW). The exact layer thickness is verified by a helium-neon triangulation laser.

Figure 3-18 shows the present machine and a detailed view of the interior of the process chamber.

Material/Build Time/Accuracy: The range of material includes the DuPont resins SOMOS 2.100, 3.100 and 5.100 as well as the EXactomer from AlliedSignal. Resins of independent producers (see Appendix A2-15) can also be used. Machine-specific build times and accuracy can also be found in Appendix A2-7.

Post-Processing: In this installation the post-processing is identical to the stereolithography-typical post-processing described in Section 3.3.1.1.

Process-Typical Follow-Up: All stereolithography-typical finishing techniques and casting processes are possible (see Section 3.3.1.1).

Figure 3-18 Laser model system (LMS). (Source: Fockele & Schwarze)

3.3.1.5 Solid Ground Curing – Cubital

Solider 4600, 5600

Cubital Ltd., Raanana, Israel

The company obviously stopped their activities in 2002. It is mentioned here, because there are still a lot of machines in the market.

Short Description: The Solider process is a stereolithography process in which the reproduction of the layer geometry and the exposure are carried out by means of a high-powered UV lamp through a mask (lamp-mask). The process of generating a mask is additional to the actual stereolithography process. The models are generated without any support constructions and they can even be placed on top of one another in the build chamber so that the machine has a very high productivity. The models are, above all, very accurate in the z-direction because the layer is milled plano after each light-exposure process. Those areas not belonging to the model are filled with wax. During the post-processing outside the machine the wax must be removed.

Range of Application: Geometric prototypes, functional prototypes (technical prototypes).

Development State: Commercialized since 1991.

History: The Israeli company Cubital, established in 1987, is one of the oldest producers. This also applies to the machine that was introduced in 1991.

Development Partners/Strategies: With the Solider, Cubital has developed a "production installation for prototypes" designed for large-scale manufacture; small single pieces cannot be produced economically. The profitable operation of this installation therefore requires the production of a corresponding number of prototypes.

The machine is not very widely used. In 1998 just about 30 machines were running worldwide, partly because the machine weighs 4 tons and belongs in a workshop with a corresponding infrastructure.

Data Formats/Software: The machine accepts STL data, even if they are not error free.

The Cubital software is considered to be effective and especially error tolerant not only when used in the Solider prototyper. It can therefore be recommended as "front end" for other systems, too. The core of the system is an algorithm that reduces the amount of data in the STL files. On the one hand, neighboring triangles on the same plane are combined by eliminating their joint borders and, on the other hand, triangles lying at a slight angle toward one another are put on one plane according to a given critical angle. An increase of the allowed tolerance of the critical angle of neighboring triangles from $0°$ to $0.05°$ alone usually results in a reduction of the amount of data by more than a third. The program known as *Solitrim* requires the *Solifile* CFL format.

The Principle of Layer Generation: Cubital derived an independent process approach from the theory of photopolymerization. The basis of solid ground curing (SGC) is the exposure of each layer by means of a lamp through a mask. In this way the time needed for the generation of one layer is to a large extent independent of the complexity of that layer. Consequently, advantages in comparison to scanner processes are always realized when large build spaces with high layer information densities need to be processed. The Cubital process (see also Fig. 3-19) combines two subprocesses: generation of the mask and generation of the layer.

Generation of the mask: From the slice program of the computer the contour data are transferred onto a glass plate by a technique similar to the Xerox^{TM}-copying process. This is done by electrostatic charge. The glass plate is then "developed." Toner adheres to the electrostatically charged areas. An opaque glass plate results with the exception of those areas to which the contour data were transferred; these remain transparent.

Generation of the layer: Through this glass mask the desired layer is completely polymerized at once by means of a lamp emitting ultraviolet light. The exposed structures, the subsequent model, is solidified; the surrounding nonexposed parts of the model stay liquid.

The liquid monomer is sucked off as completely as possible in the next process step. The entire model is then covered by a layer of liquid wax, whereby especially the lower parts of the model, where the liquid monomer has been removed, are completely filled with wax. The wax, which prior to application is liquified, is cooled down afterwards to about freezing temperature. It solidifies thereby sufficiently to enable the entire model surface to be trimmed by surface milling in the next process step. A new layer of resin is applied by an active coating system and the exposure process starts again. The preceding mask is cleaned by removing the toner and a new mask with the copy of the next layer is generated.

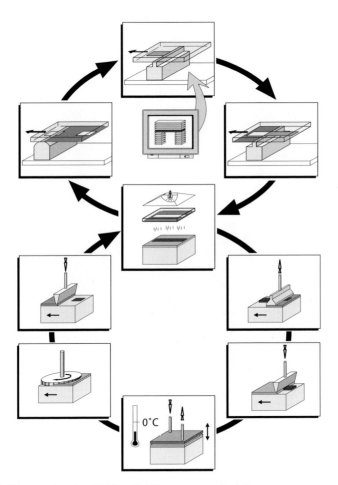

Figure 3-19 Solid ground curing (SGC) – Solider-process, principle

Since the wax that has not been removed by milling stays in the model, it serves as supports for the subsequent model parts. The Cubital process therefore can generate any complex 3D structures without additional supports.

Care must be taken that superfluous wax is completely sucked off during the milling. If some resin stays in the model the exposure of the next layer will show muddy surfaces and blurred contours. Problems such as inaccurate exposures can also occur if the masks are optically not completely sealed.

System Type/Construction: The Cubital Solider System is very complicated in regard to process technology. The machines therefore are correspondingly complex. Trimming by surface milling especially requires a solid machine construction. The Cubital machine (Solider 5600) resembles a machine tool in design, measurements (L · W · H = 4.1 · 1.7 · 1.5m), and weight (ca. 4.5 t). Therefore it is essential to operate this machine in a proper workshop under continual supervision. The main components are arranged on a common machine bed in such a way that they are positioned over the build chamber for each process step. Figure 3-20 shows a Cubital machine (Solider 5600) with the casing cover is removed. A 4 kW lamp serves as a light source.

Figure 3-20 Solider 5600 prototyper. (Source: Cubital)

The ability to generate models without further supports has the advantage that a machine can be filled with any number of interpenetrating models. Also, model parts that are at first not connected with one another (islands) can be produced independently. They need neither any supports nor bases because the wax keeps them in their correct position until they are connected with the rest of the model. The Cubital process can also generate multipart models with movable elements. The machine is most advantageous when many models (up to 10, depending on their sizes) are generated simultaneously and when their walls are relatively thick. A Solider 5600 has the capacity of about five SLA5000 and 20 to 25 SLA250/50. Even though the productivity of the SLA7000 is four times higher than that of the

SLA5000 according to the producer, the productivity of the Cubital machine is twice as high as that of the 3D Systems machine according to the documentation.

Material/Build Time/Accuracy: Figure 3-21 shows some typical and especially large thin-walled models for which the Solider process is especially suited.

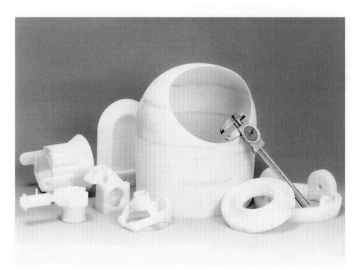

Figure 3-21 Solid ground curing (SGC), typical models, (Source: Cubital)

Post-Processing: Post-processing, depending on the construction principle, is fundamentally different from that of laser scanner based stereolithography processes.

After completion of the build process the wax must be removed from the model either by melting or by washing the model in citric acid. This is done in special installations called "dewaxer" stations. Because it is not possible to remove the monomer completely during the sucking process the wax is in fact a wax-monomer-solvent mixture that must be disposed of after the dewaxing.

Process-Typical Follow-Up: All stereolithography-typical finishing techniques and casting processes are possible (see Section 3.3.1.1).

3.3.1.6 Micro-stereolithography – MicroTEC

RMPD (Rapid Micro Product Development)-CIM

MicroTEC GmbH, Duisburg, Germany

Short Description: A Prototyper (RMPD-CIM) is available for the application of different micro production (RMPD) processes either working point by point (RMPD, RMPD-stick2) or using masks (RMPD-mask). The machines work according to the principle of laser stereolithography. The parts can be building "stand alone" or stuck to substrates. The maximum size of the component is 50 · 50 · 50 mm (approx. 2 · 2 · 2 in.); a definition of < 10 μm can be achieved. The production of functional microsystems is facilitated by the possibility of putting parts such as jewel bearings, shafts, and so forth into the cavities during the build process or to join different materials. RMPD-nanoface is another RMPD system for high surface qualities in the sub-nanometer range.

Range of Application: Geometric prototypes, functional prototypes, technical prototypes, short run and mass production.

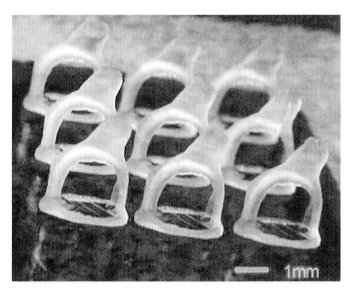

Figure 3-22   Micro stereolithography: Ear bone stapes produced in parallel (RMPD). (Source: MicroTEC)

Development State: Models and micromechanical devices are produced for customers.

History: In 1996 MicroTEC was founded R. Götzen at Duisburg, Germany. Although the market introduction of the machine family had been announced in 1999, which indicated, that MicroTEC originally wanted to sell it's equipment, they now act as a service bureau.

Development Partners/Strategies: The company offers customer specific services, short run and mass production of microsystems and micro structured parts. The name itself, "Rapid Micro Product Development," underlines this strategic approach. A corresponding modularization is clearly recognizable. Scanner and mask processes fulfil various requirements for accuracy. Microoptic controlled direct contouring or multi mask applications, enable the production of various numbers of pieces.

Data Formats/Software: The machine adopts geometric information directly from the CAD via neutral interfaces. No information is known about the types of interfaces. An automatic slice program is supplied by the producer.

The Principle of Layer Generation: The entire solidification process follows the principle of stereolithography. The shaping of the parts is done with a pulsed laser either point by point (RMPD) or using masks (RMPD-MASK). The simultaneous exposure of many parts is effected point by point to achieve any desired shape using a array of micro optics. If the part is stuck to a substrate, e.g. a wafer, the process is called RMPD-stick2.

System Type/Construction: How the growth in steps of less than 1 μm in the z-direction is realized remains as much a secret as the method that allows micro-elements such as shafts to be inserted and, after their insertion, to find the exact current z-coordinate while avoiding contamination by process foreign material. It is imaginable, but this is pure speculation, that a glass plate serves as the upper (or lower?) cover of the build chamber which would enable an exact z-positioning and an absolutely flat x-y plane. That still leaves the secret of how the resin is removed from the glass plate after polymerization. After all, in the beginnings of stereolithography, models that had to be exceptionally precise were built on glass. The notion that the component lies in the resin bath as in classical stereolithography processes is barely credible because the component must be introduced into the resin and a cleaning process is necessary.

Nothing is known about the way how the masks for the RMPD-MASK process are produced.

Material/Build Time/Accuracy: Acrylates and epoxy resins are used. According to producer information various materials may be used for one component; different mechanical-technological properties are thereby achieved depending on their area of usage.

The construction of a forked pipe (micro-mixer) of approx. 900 μm in height with three joints of ca. 300 μm diameter and a wall thickness of not quite 100 μm takes about 4h. The average layer thickness is ca. 10 μm. The resolution of a layer thickness can be less than 1 μm and that in x-y direction less than 10 μm.

Post-Processing: Nothing is known about post-processing probably because this would allow conclusions to be drawn about the building process.

Process-Typical follow-up: It is possible to strengthen the surface by metalizing it or by coating it, with diamonds, for example. All duplicating processes suitable for micro parts are applicable.

Further information to micro stereolithography and related procedures, see [VAR01].

3.3.1.7 Stereolithography – EnvisionTechnologies

Perfectory Standard, Perfectory Mini

Envision Technologies GmbH, Marl, Germany

Short Description: EnvisionTec's prototyper works with a upside down arranged Stereolithography process. The bottom of the barrel shaped machine body contains a DLP (Digital Light Processing Technology – Hewlet Packard / Digital Micromirror Device . DMD) which directly projects each cross section onto the build area on the top. The build space and material container is basically a glass device with the upper side open, mounted on the top of the machine body. The layer surface is defined by the down side glass plate of the material container. Each layer is cured at once though the glass plate by the light emitted from the DLP. A upside down arranged elevator dips it's elevator plate from the top side into the material container. The distance between the elevator plate and the glass plate defines the thickness of the first layer. After curing, the first layer sticks to the elevator plate while a special coating prevents it from sticking to the glass plate. The elevator then is moved by the amount of one layer thickness allowing the uncured resin to cover the space that marks the volume of the next layer. The model is made layer by layer the same way, growing upside down out of the material container.

Strategies: The PERFACTORY is a easy to handle mobile stand alone device that needs no re-installation after moving. It is specialized to build rather small detailed parts preferable for the jewelry industry and for medical applications. The manually exchangeable material container allows a quick exchange of material. It can be refilled automatically or manually. Because of the small vat and the small parts to be build, only a little amount of material is needed.

Development State: The machine was presented at the 2001 Euromold fair. The company runs programs with test clients

The machine is designed as a low cost system in terms of investment and costs for spare parts.

Data Formats/Software: The company offers a easy to handle Rapid Prototyping software. It takes STL or CLI data. The software offers Auto Mode and Expert Mode: in Automatic Mode, the system is runs automatically. In Expert Mode process parameters can be set up manually, after processing the 3D-CAD data using RP Software (Marcam Engineering, Materialise).

The Principle of Layer Generation: Each cross section is projected in total on the bottom of the glass plate by a DLP projector equipped with a high-pressure mercury vapor lamp. The whole layer is solidified at once. The PERFACTORY models need supports.

System Type/Construction: The prototyper consists of a barrel shaped body, a vat made of glass and a movable platform that can be controlled in z-direction by a high precision drive. The body contains a DLP projector, the control unit with a display on the top and the material feeding device. The change of material is very simple and quick with material supply cartridge. The material feeding device is equipped with a tube pump which allows the whole tube to be exchanged without cleaning. In the manual mode the feeding system is passed by and the resin is put in manually. The vat and the elevator are covered by a transparent UV –blocking tube that gives the machine it's special appearance.

Material/Build Time/Accuracy: Part of the company's philosophy is the development of their own high sensitive resins. The first material available is a redish methacrylate. A variety of materials in different colors and with different mechanical properties are currently under development.

The building speed in z-direction is approx. 10 inches per hour which makes the machine very fast. The DLP projector delivers a x-y grid of 1024 · 768 pixels. Because of the mirrors, each pixel ranges from 139 · 139µm to 249 · 249 µm (PERFACTORY Standard) and from 41 · 41 µm to 73 · 73 µm (PERFACTORY Mini) depending on the build space. The layer thickness can be varied form 30 to 80 µm (Mini) and from 50 to 150 µm (Standard).

The Perfactory Standard achieves a x-y resolution of 180 DPI with 142 · 106 mm (5.6 · 4.2 in.) build area and 100 DPI with 255 · 191 mm (10 · 6.4 in.) build area. The maximum build height (z) is 250 mm (9.8 in.).

The Perfactory Mini achieves up to 800 DPI resolution with a x-y build plane of 42 · 31 mm (1.65 in.) and 1000 DPI in z-direction with a max. part height of 50 mm (1.96 in.).

Post-Processing: The supports need to be removed and the model must be cleaned. As an advantage the models can be cleaned with alcohol-free solvents.

Process-Typical Follow-Up: All follow-up processes suitable for stereolithography parts are applicable. Until now there are no specific experiences with vacuum casting and indirect tooling processes.

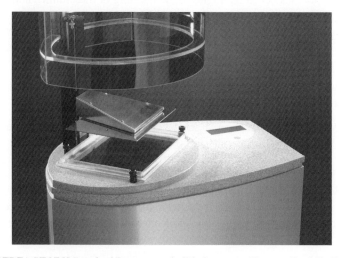

Figure 3-23 PERFACTORY Standard Prototyper, build plane area (Source: EnvisionTechnologies)

Figure 3-24 PERFACTORY Model "bicycle" (Source: EnvisionTechnologies)

3.3.1.8 Stereolithography – PolyJet – Objet

Quadra, Quadra Tempo.

Objet Geoemetries Inc., Rehovot, Israel
Objet Geometries, Inc, Mountainside, New Jersey, USA

Short Description: The Objet prototyper is a multi nozzle print head system (named PolyJet) that prints liquid photopolymer directly on the part and exposes it to UV light simultaneously. The machine basically is a "3D printing type" device, while the solidification of each layer is a stereolithography process.

Development State: Commercialized since 2001.

History: Objet was Founded in 1998 by Scitex Corporation and others. The Quadra Tempo was announced in 2002.

Development Partners/Strategies: The Objet technology was born out of the desire to apply experience in graphic ink-jet technologies to the building of three-dimensional models. The Quadra prototypers are used as three-dimensional network printers for digital files created by designers and engineers in a wide variety of industries who need to physically verify their designs. The company develops and uses patented proprietary hardware, software and polymer materials.

Data Formats/Software: The system takes STL files from solid 3D models which are processed using the Objet Studio software.

The Principle of Layer Generation: Each layer is printed by jetting model and support material onto the build platform. The multi nozzle printhead that contains 1536 nozzles allows to print the whole width of each layer at once. Each layer is cured (hardened) by the exposure of UV light from two lamps located on the inkjet head assembly. For models with undercuts and overhangs, the Objet Quadra deposits support material which is also a photopolymer. The supports are solids that completely fill in undercuts and any type of hollow structures. Due to this, always the whole circumscribed volume of the part has to be build, partly with building material, partly with support material, influencing the material consumption.

System Type/Construction: The Quadra is a stand alone system with a build space of 270 · 300 · 200 mm (10.6 · 11.8 · 7.8 in.). The gantry which contains the print head and the lamps, moves in x-direction. The build platform repositions in y-direction if the part is wider than the printhead. The platform also moves in z-direction after each layer is deposited. Especially the Quadra Tempo is designed for unattended long-term runs. It is a advanced version of the Quadra and allows to speed up the building process by approx. 20% to 25 %.

Material/Build Time/Accuracy: The Quadra prototypers use a proprietary photopolymer (called FullCure M-510T-Y) for the part structures which is a acrylate based polymer. For the supports a curable water soluble polymer is used. No properties are published. The layer thickness is 20 microns (0.0008 in.). A print resolution of: x = 600 DPI, y = 300 DPI and z = 1270 DPI can be achieved. PolyJet in a pulished case study provided a higher accuracy and a smoother surface than SL and SLS [GRI02].

Post-Processing: The process requires no post curing but the removal of the supports. The gel-like support material is designed for easy removal by mechanical means or by water jet.

Process-Typical Follow-Up: All follow-up processes suitable for stereolithography models are applicable.

Figure 3-25   Quadra prototyper (Source: Objet Geometries)

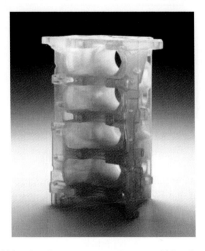

Figure 3-26   PolyJet part build by Quadra prototyper (Source: Objet Geometries)

3.3.2   Laser Sintering

3.3.2.1 Machine-Specific Basic Principles

The Principle of Layer Generation: In the laser sintering process (see also Section 2.3.2.1 and [NOE97]) particles of usually 50 to 100 µm which are closely packed together to form a powder bed are slightly pressed if this is necessary for the process. They are then melted locally by a laser beam, solidified by cooling due to heat conduction and are thereby joined together to form a firm layer (Fig. 3-27). By lowering this layer and recoating it with powder in analogy to the first layer, the second layer is solidified and connected with the first.

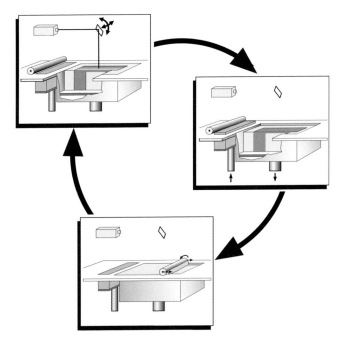

Figure 3-27 Laser sintering process, principle

Design: It is important for the technical functioning of this process to construct the process chamber in such a way that it can be preheated up to very nearly melting temperature of the sinter material. The laser beam therefore only needs to add a small differential energy for the sintering. The process temperature must be kept as steady as possible and within narrow tolerances (a few degrees). Furthermore, oxidation of the material must be avoided, which is usually achieved by inerting the machine. This is done by generating a nitrogen atmosphere inside the machine (0.1 % to 3.5 % residual oxygen, depending on the material).

The laser sintering process always works without supports because the unsintered powder stays in the bed and supports the model. Depending on the geometry of the model and the material being used it has proved to be useful to build a platform (base) as well and to build on this base. In addition or alternatively, firm understructures can be used if required. Starting with about half laser performance and double scanning speed, and by adjusting the laser performance as well as the scanning speed to the optimal parameters within several layers, structures with low distortion are generated that favor a reliable model construction.

To maintain an even temperature field, it is useful in some cases to place loose parts deliberately near the model or to put a grid around the component.

Some users suggest positioning the model, especially long drawn-out parts, under the so-called Kodak angle relative to the cylinder's longitudinal axis. It is assumed that especially distortion-free components can be generated under this angle. Some sources mention an angle of about 15°, others about 10°.

In practice nearly all models are positioned in the machine without a base according to criteria of accuracy and economy. When the build parameters are carefully matched, and especially when the temperature fields are kept even, the results are excellent.

Material: Plastic materials are best suited to laser sintering owing to their relatively low melting temperatures and their low heat conduction properties which facilitate a local limitation of the sintering process. The process can, however, be used in principle with all materials that can be incipiently melted and sinterfused under thermal treatment and that solidify again after cooling. The term thermoplast, usually used only for plastic material, must therefore be understood as a generalization.

Particles of 20 μm up to about 100 μm diameter are used in sintering machines. By sifting fractions are achieved with grains that are usually 90% below the given diameter. Correspondingly rough surfaces are the result. This is especially the case with amorphous materials (polystyrene, polycarbonate) which tend to be molten only incipiently and therefore essentially retain their form. Crystalline materials (polyamide) that become completely liquid form smoother surfaces owing to their surface tension. They tend to form dimples, however, so that rough surfaces result here as well.

Because the sintering process is not a pressurized process; the achievable density is lower than with injection-molded materials. This applies especially to amorphous materials that have relatively large hollow spaces and therefore reach densities of only 60% to 85 %. These properties which act negatively on the endurable load of the parts, are beneficial toward parts that are melted out in a precision casting process. Amorphous materials are more easily melted out and are not as susceptible to "shell cracking." Crystalline materials can easily reach densities that are very near those of injection-molded parts.

There are differences also in the shrinkage of crystalline and amorphous materials. Crystalline materials alter their volume during the melting process quite significantly and therefore tend to shrink. Amorphous materials are considerably less prone to shrinkage depending on the density achieved. The shrinkage usually leads to curling, the tendency of the models to "roll up" in the direction of the heat source. This tendency can in principle be controlled by shrinkage factors which should already be taken into account during the generation of the build data. As these factors are empiric values they cannot be defined with sufficient accuracy right from the start.

Build Time/Accuracy: Build times and accuracy depend on the geometry and therefore on the positioning of the component in the build chamber.

The build time in the x-y-direction is much higher than that in the z-direction. Typical build times are approx. up to 10 mm/h. The accuracy are limited by the laser beam diameter. Beam diameters of about 0.4 mm result in accuracy of ±0.15 up to 0.2 mm and finest details such as freestanding walls of 0.5 mm thickness minimum.

Post-Processing: After the build process is finished the model is completely enclosed in a powder cake. Most sinter processes take place at a temperature of between 170 °C and 200 °C, especially when plastic materials are sintered. To ensure a uniform cooling it is important that after the last layer an additional layer of powder several centimeters thick is applied. After the compound has cooled completely, which may take several hours owing to the poor heat conduction, the powder cake is carefully removed from the outside (Fig. 3-28). Although, in theory, the compound is merely embedded in a loose powder cake and only needs to be taken out and the powder blown off, in practice it is wise to proceed carefully. First, the compound can easily be damaged because its position is not exactly known; second, there are, depending on the temperature control, slightly sintered areas especially around the model (fleeces) that have to be removed with great care by using special tools. Therefore, it demands patience and skill to clean sinter models especially those with internal hollow spaces, drillings, and fine details. To exacerbate the situation models and powder have the same color.

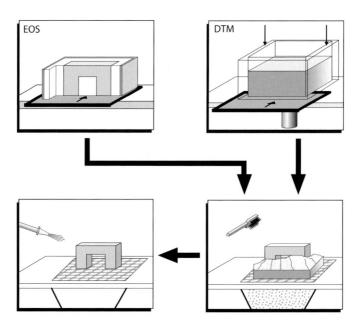

Figure 3-28 Laser sintering, Post-processing, principle

After the parts are cleaned in a so-called post-processing, the surface is treated further by manual polishing or sand blasting. Parts of models or broken off parts of models can be glued on with cyanoacrylate glues or with epoxy resins. To finish the models drums with polishing material can also be used. It has to be taken into account that the type of abrasive used determines the amount of material removed. There is a high risk of rounding sharp-edged corners. Because sinter models are generally porous, all infiltrating surface sealings may be used. This includes all kinds of hard wax, epoxy resins, and also primers on an enamel base.

Follow-Up Processes: Laser sintering processes are preferably used for functional prototypes (functional parts). Therefore, the direct application is used more often than casting processes. Vacuum casting is possible in general but requires that the surfaces be exceedingly well finished.

It is especially interesting to use polystyrene or polycabonate models directly for precision casting. This procedure is successful only if an appropriately careful wax impregnation and surface preparation is applied. The model itself must be pre-heated up to approximately the temperature of liquid wax (between 190 °C and 210 °C). The impregnation can take up to 30 min; if critical spots have to be reworked manually it could take significantly longer. Finally the classical precision casting process is started. The burn out should be done with sufficient oxygen so that the chemical reaction is supported. A ventilator is probably helpful.

In areas with thick walls ash can accumulate which is difficult to remove later. Such material accumulations can even cause local shell crackings. A number of precision casting foundries, however, have, in cooperation with the model makers, managed to control this process so reliably that for every polycarbonate model presented, a casting model is returned. The advent of CastForm PS (3D Systems) material should solve these problems once and for all.

Following the basic principles of laser sintering, machines have been designed that differ especially in details, target applications, and consequently in the materials used. At first some of the constructive solutions were devised owing to the patent situation but later prevailed and blazed the way forward.

3.3.2.2 Selective Laser Sintering – 3D Systems / DTM

Sinterstation 2500^{plus}, Vanguard

3D Systems Inc., Valencia, CA, USA

Short Description: Family of prototypers for the sintering of plastics, metals, and sands. The machine works on the laser scanner principle. A CO_2 laser scans the surface of a powder bed and incipiently melts the particles that form the model after solidification. Owing to the process technique no supports are necessary.

Range of Application:

Plastics: Geometrical prototypes, functional prototypes (technical prototypes)

Metals: Technical prototypes, direct tooling (series parts via tooling)

Sands: Series parts.

Development State: Commercialized since 1992. An estimated number of over 300 machines have been sold worldwide.

History: 3D-Systems sintering activities go back to the DTM corporation of Austin, TX. The DTM Corporation was established in 1987 by Carl Decker as a branch of B. F. Goodrich and only at the beginning of 1999 was it sold to a group of investors. The laser sintering process is called selective laser sintering (SLS) by DTM and is protected by patent (Carl Decker). The first machine, the Sinterstation 125, was introduced in 1989/90. For some time it was used in the service office (from 1990 onwards). The first industrial machine for SLS, Sinterstation 2000, was developed simultaneously. Marketing began in the United States in 1992 and in Europe in 1993. The largely identical Sinterstation 2500 with a rectangular build chamber followed in 1997; the current Sinterstation 2500^{plus} was introduced in 1998. The Vanguard (2001) is based on the 2500^{plus} hardware with some specific impovements such as scanners, heating, and control. In additon the software was upgraded. In 2001 DTM was taken over by 3D Systems.

Development Partners/Strategies: 3D/DTM wants to be able to use every kind of material in one prototyper. Software-material packages have been developed that are matched to one another. The development partner is ISG, St. Gallen (Switzerland). The developers refrained from optimizing the machine for one special class of material, thereby enabling the user to choose freely from various materials. In practice it depends very much on the user as to whether he takes advantage of this. The expenses for the software material packages, the time needed to reset the machine, and the necessary training mean that in practice most users only run one class of material.

3D's present objective is to break into the market for plastic functional parts and, strategically, to enter the market for so-called "functional metal parts." These are primarily complex series-identical or series-close model parts that can be tested for optimization of (steel) casting components under series conditions (see also Section 5.2.1).

Data Formats/Software: The machine accepts error-free and complete STL data. Neutral interface formats can also be read by means of an additional module.

The standard program system supplied by the producer is similar to the MAGICS RP Software from Materialize and supports the most important operations.

The Principle of Layer Generation: The scan strategy consists of a raster scanning of the relevant cross section (raster technique). This results by necessity in stair steps in the build plane also, which further diminish the quality of the surface. The stair-stepping effect manifests itself when the model is generated without any additional borders. To make the stair-stepping effect less noticeable the scan direction is turned after each layer generation, thereby spreading the steps more evenly over the model. If this should be avoided (especially for thin walls) scan strategies which run along the contours first and then raster scan the internal areas are preferable. To avoid thermal imbalances the scanning takes place alternately in the +y and -y direction, that is, from the "front" or from the "back."All in all the machine control is organized in such a way that the computer calculates one layer at a time while the machine is building the preceding one. This process is effective as long as the calculation of one layer takes no longer than the build of this layer. Otherwise the machine has to wait for the slice algorithm. This can easily happen with extensive data records and complicated geometries such as, for example, the cranial base of a medical cranium model.

System Type/Construction: The process chamber consists of two reservoirs and two overflow containers and between them is fitted the build chamber with the powder bed. The Sinterstation 2500^{plus} has its material containers above the powder bed surface. The material is transported by a feed roller from the supply vessel onto the powder bed. Superfluous material passes to the material container on the opposite side, or to the overflow. Those two material containers allow a bidirectional and therefore faster recoating after one layer is finished.

The build chamber is fitted with an adjustable piston that is lowered by one layer thickness after each layer generation is finished and before the layer is recoated with powder. The build chamber was originally circular (Sinterstation 2000) and is now rectangular (Sinterstation 2500/2500^{plus}). The entire machine is preheated to a temperature of about 4 °C lower than the melting temperature of the applied construction material. The machine is then flooded with nitrogen. A continuous nitrogen stream through the powder bed supports the even temperature distribution in the build space (down draft). Figure 3-29 shows the principle design of the DTM Sinterstation 2000^{TM}. The design of the Sinterstation 2500 and 2500^{plus} are principally identical apart from the build chamber which is now rectangular.

Laser and scanner system are separated from the build chamber. The laser beam (CO_2 laser) reaches the build chamber through a special window [zinc selenite (ZnS)] that can be removed when the machine is cleaned. The nitrogen maintains an even temperature because it is preheated and then cooled while it flows around the window, thereby keeping it clean. A large window protects the build chamber on the outside and allows the build progress to be observed.

To avoid the curling effect, it is important that the model is properly oriented in the build chamber and that the machine is warmed up thoroughly.

To obtain perfect models it is very important to thoroughly adjust and monitor the temperature and temperature distribution in the build chamber. Temperature variances of only a few degrees can result in the creation of unuseable models, which are either badly sintered or, due to too high a temperature, severely curled or, depending on the material, even of a different color. The temperature dynamics therefore play an important role. This is why a much lighter and thermally faster machine than the Sinterstation 2000 was designed. Optimized heating from above the powder bed and the heating of the upper wall areas of the build cylinder together with a fast control provides even temperature fields. The knowledge that material mass is not only thermally stable but also thermally inert and therefore difficult to control led to the Sinterstation 2500plus being an independent machine and incompatible with its predecessors. In future it will be the only platform supported by 3D Systems.

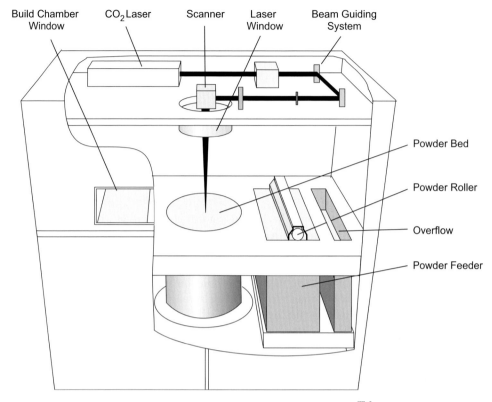

Build Chamber   CO$_2$ Laser   Scanner   Laser   Beam Guiding
Window   Window   System

Powder Bed

Powder Roller

Overflow

Powder Feeder

Figure 3-29 Principle design of a sinter machine (DTM Sinterstation 2000^{TM})

It is also advisable to ensure that the heated roller is roughened at longer intervals to achieve an optimal recoating.

The nitrogen supply is introduced through a built-in air separation installation. This ensures that the supply of nitrogen does not expire during the build process. If this should happen the model-making process would have to be terminated and the model would be unuseable.

Before the completed model is removed, the entire machine has to cool down. This process could take several hours depending on the material. Consequently the entire machine needs to be warmed up again before a new build process is started. These periods are clearly part of the build process, but they must be included in the model-making time and accounted for in the costing calculation. In Benchmark tests they are sometimes ignored.

Owing to the design, the entire machine has to be exposed to the normal oxygen atmosphere when the models are removed. For a new build process, therefore, a completely new inert atmosphere must be reestablished.

The machine is calibrated by exposing a special foil with markings. The interpretation is carried out by the producer in the Uniteed States. The process is protracted.

Figure 3-30 shows the Sinterstation 2000. The doors are opened so that both material containers and overflow containers can be seen clearly in the lower part of the machine. The circular border of the build cylinder is visible in the middle of the actual build platform. On the far left, the coating roller can be seen. Concentrically above the build chamber is the protective cover of the laser beam which is closed off by a ZnS window. In a separate part of the machine on the right, the controlling computer and the operating panel can be seen.

(a)   (b)

Figure 3-30   Sinterstation 2000: exterior view (left) and interior view with opened doors (right).
(Source: DTM)

Figure 3-31 shows the exterior view of the 3D Vanguard Sinterstation.

Figure 3-31   3D Vanguard Sinterstation, exterior view. (Source: 3D Systems)

Material/Build Time/Accuracy:

Material – Plastic Materials

Under the name DuraForm two polyamides (approx. equivalent PA11) are available for the Sinterstation 2500/2500^{plus}, either as a pure nylon or glass-filled under the name DuraForm GF. The grain size is typically 90% < 50 μm and is equivalent to the former fine nylon. Glass fibers cannot be used because of the process, but glass globules may be mixed into the powder. They impede extension thereby increasing the mechanical stability. The REM photo (Fig. 3-32) shows that, due to accumulations of globules, internal hollow spaces are formed that influence the dynamic properties of the material; test results are therefore not necessarily applicable to series parts.

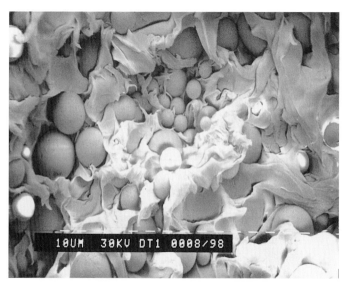

Figure 3-32 REM photo of glass-filled polyamide. (Source: Conrads/FH Aachen)

For elastic models, a thermo-plastic elastomer is available known as DMS-Somos (Somos 201TPE, DuPont). It is used for the simulation of rubber-elastic materials. With the same shore hardness series properties such as tightness and locking capability can be tested at room temperature. The behavior of the materials differs considerably between low temperatures (< -30 °C) and high temperatures (> +50 °C) as well as over long periods of time.

TrueForm PM (TrueForm II) on a polystyrene basis has replaced polycarbonate as the material for precision casting. It was introduced in early 1996 as a smooth surface pattern making material named TrueForm (I). Its surface properties are far superior to those of fine polyamide. It is basically a copolymer of the acrylate type that cannot be polymerized further. The thermal shrinkage can be compared with that of acrylates. TrueForm I has proved a success in the United States, but it seems not to have satisfied the higher demands of European foundries. TrueForm PM, also known as TrueForm II, has been improved with regard to its reproduction and melt-out properties. Since the summer of 1999, DTM supplies an improved casting material known as CastForm PS, an improved polystyrene, which the company claims to be suitable for precision casting of aluminum and titanium, and also of low melting point alloys. It has already replaced TrueForm PM.

Material – Metals

DTM has developed a multistep indirect metal process (RapidTool) on the basis of polymer-coated metal powders (RapidSteel 3.0) for the production of metal models and molds. This process is based on the knowledge that all existing sintering machines can sinter any material if it is embedded in plastic [MAR91]. On this basis a multistep metal process – the so-called indirect metal process – was developed:

First the model is sintered in a prototyper from plastic-coated metal powder. The model achieves a "green stability" (green product) owing to the plastic coatings of the metal particles that sinter fuse with one another.

This model is then resintered in a reducing hydrogen atmosphere inside a special high-temperature oven, thereby expelling the plastic material also. This is especially important as in this way oxide skins, which would considerably lower the wettability of the rather loosely sintered metal particles (brown part), are avoided.

In a third process step, a low-melting metal, usually copper, is infiltrated into the hollow spaces of the porous model. This is done in the same oven if the infiltrate has a corresponding melting temperature. The infiltrate should merely be placed right next to the model to enable the infiltration to take place by capillary forces.

Figure 3-33a shows such a brown part before infiltration, and the copper cylinders used for infiltration. Figure 3-33b shows the result after the infiltration has taken place.

(a)

(a)

Figure 3-33 Indirect metal process. (a) Green part (before infiltration). (b) Brown part (after infiltration). (Source: DTM)

A metal model results that approximately possesses the properties of aluminum and is especially suited for the production of injection molded parts in small numbers and of low process parameters [ARG91].

Figure 3-34 shows some representative results of mold making after mechanical finishing.

Figure 3-34 Indirect metal process: Injection mold, parts. (Source: DTM)

This is a relatively complicated process that demands a great deal of experience and additional investment. Furthermore, the indirect metal process has the paramount disadvantage that, owing to the multiple heat treatment via green product and brown part, the shrinkage factors which cannot be disregarded, become so complicated that it is impossible to control them satisfactorily without empiric values. In addition this process is very time consuming.

Copper polyamide (PA) is a simple to process but also softer material that is used preferably for molds and tools for the casting of soft materials.

For the Sinterstation 2000[plus] a build speed of approx 11.5 mm/h is reported. It is, therefore, about 1.6 times faster than the Sinterstation 2000 and about 1.3 times faster than the Sinterstation 2500. The accuracy are claimed to be in 90% of the cases less than ± 0.12 mm and are therefore three times better than those of the Sinterstation 2000. An installation known as Vector Bloom Elimination (VBE) Circuitry also has an influence on this result. This installation controls the laser power adjustment and helps to avoid an undesired widening of the laser beam, especially at the end of the vectors and at large excursions.

It should be remembered that, depending on the material used, the useable space in the build chamber can easily become smaller owing to specified minimum distances to be kept between walls and model.

Under the names SandForm Si (on a silicate basis) and SandForm ZR (on a zirconium basis) two qualities of polymeric bonded sands are made available which can directly be sintered to cores for sand casting, thereby producing series-identical sand casting prototypes.

Build times and accuracy typical for laser sintering are achievable (see also Section 3.3.2.1).

Post-Processing: The post-processing is identical to the post-processing typical for laser sintering as described in Section 3.3.2.1.

After the actual build process is completed the entire machine must be cooled down, which could take several hours. Only then should the whole cake with the embedded component be taken out. After the model has cooled down sufficiently it can be broken out of the powder cake and cleaned.

Process-Typical Follow-Up: All follow-up processes typical for laser sintering can be applied (see also Section 3.3.2.1). It has to be taken into account, however, that vacuum casting does not result in surfaces as smooth as those achieved by stereolithography.

3.3.2.3 Laser Sintering – EOS

EOSINT P350, P700, EOSINT M250 Xtended, EOSINT S700

Electro Optical Systems GmbH (EOS), Krailling-Munich, Germany

Short Description: Family of prototypers for the sintering of plastics, metals, and sands. The machine works on the laser scanner principle. A CO_2-laser scans the surface of a powder bed and incipiently melts the particles that form the models after solidification. The process requires no additional supports.

Specially optimized prototypers exist for the material groups plastics (P), metals (M), and sands (S).

Range of Application:

Plastics: (EOSINT P) geometrical prototypes, functional prototypes (technical proto-types), series parts by means of precision casting.

Metals: (EOSINT M) technical prototypes, direct tooling (series parts by means of tooling)

Sands: (EOSINT S) series parts (castings)

Development State: Commercialized since 1994 (P), since 1995 (M, S).

History: Shortly after developing its first stereolithography machine in 1991, EOS presented a sinter machine for sintering plastics on which the later EOSINT model range was based. The EOSINT P (P350/60; now P350 and P700 with two lasers), a machine for sintering polystyrene and polyamide was followed by the EOSINT M (M160 (Prototyp)/M250/M250Xtended), a machine for the direct sintering of metal alloys and metal powders. The P350 was developed into the S350 as prototype of an installation for sintering casting sands and later into the EOSINT S700 for the production of (preferably) cores for sand castings.

Until EOS gave up the stereolithography business it was the only company worldwide that built both stereolithography and sinter machines at the same time.

Development Partners/Strategies: EOS relies on the optimization of a machine-material system and on an accordingly specialized prototyper. The production line P is optimized for the processing of plastics, M for metals, and S for sands. The performance of the laser, the scanner and the scan strategy match the material in hand. Details such as nitrogen generators are brought into play only if the material requires them. EOS is continually searching for development partnerships with material producers such as Electrolux (Finland) and software suppliers.

All machines work with CO_2 lasers. The lasers of the P line have a capacity of 50 W, those of the M line 200 W or 250 W, and those of the S700 work with two laser scanner-units with one 50 W CO_2 laser each.

Data Formats/Software: The machines accept complete STL and contour data (CLI, Common Layer Interface).

EOS supplies a special rapid prototyping software known as RP Tools. The installation also accepts common independent rapid prototyping software.

The Principle of Layer Generation: The principle of layer generation is identical to those processes typical for laser sintering as described in Section 3.3.2.1.

The scan strategy is adjusted to the respective material.

System Type/Construction:

EOSINT P – Plastics Sinter Machine

The EOSINT-P350 (Fig. 3-35) has a rectangular build chamber with approximate dimensions of 350 · 350 · 600 mm. The build chamber is circumscribed by a cutout in the mounting plate of the machine marking the build plane. Beneath this is a freely movable platform that matches the cutout. The EOSINT-P700 (Fig. 3-36) with two laser scanner system is basicly alike but shows a much bigger build chamber and a higher productivity. The size of the build chamber is 700 · 380 · 380 mm which is sufficent to match most automotive parts, especially those who are related to engine development e.g. cylinder heads.

The EOS-machines work with the laser scanner principle, but have special constructinal features depending on the material used. One of them is the recoating system. (Fig. 3-37).

This consists of a channel-like open powder feeding unit, converging toward the powder bed, that travels over the powder bed as a ruler and in this way applies the necessary amount of powder. The entire unit vibrates to optimize the flow properties of the plastics powder. The coating system is made in such a way that the powder bed is deliberately slightly precompressed.

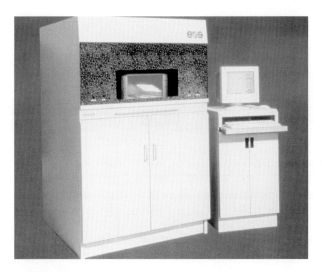

Figure 3-35 Plastics sinter machine EOSINT P350. (Source: EOS)

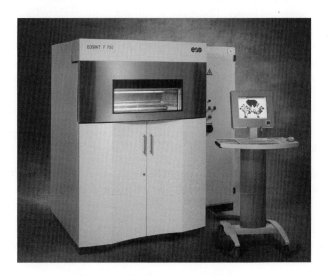

Figure 3-36 Plastics sinter machine EOSINT P700. (Source: EOS)

The necessary borders for the powder bed are also built during the build process. Although this takes time and material, it enables the complete models to be removed from the machine while still embedded in these walls and the supporting powder. The machine is designed in such a way that it is not necessary to wait until it is cooled down completely, which could take several hours owing to the low heat conductivity of plastics. The model embedded in the powder cake can be removed from the machine while still hot. The recharging time between two build processes is thereby shortenend.

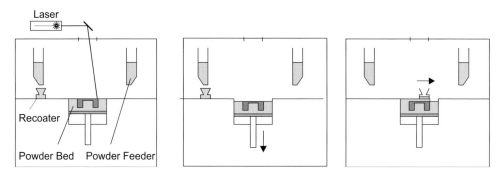

Figure 3-37 Schematic of the recoating system for EOSINT P prototypers

The EOS machine is optionally fitted with an integrated nitrogen generator that renders the user independent of gas supplies.

Machine calibration: The system must be optimally calibrated to obtain the full benefit of the machine's capacity. The system known as Optical Scanner Calibration leads the laser beam over thermosensitive paper. The result is subsequently analyzed outside the machine by means of image processing and a suitable calibration is suggested. It takes between 15 and 45 min to calibrate the machine, depending on grid dimension (usually 5 mm) and machine size. The analysis together with the subsequent calibration will take several hours if it can be done in-house. If the producer has to analyze the result it will take longer.

EOSINT M250Xtended – Direct Metal Laser Sintering (DMLS) – DirectTool

The EOSINT M line was developed especially for the sintering of metal powders with the DMLS – Direct Metal Laser sintering process.

The EOSINT M250Xtended machine is optimized toward DMLS technology for the production of mold inserts for plastic injection castings and metal pressure diecasting. This process is called DirectTool.

In addition to the transfer of geometric data the software allows the transmission of reference points that facilitate a reliable and effective finishing with conventional machine tools.

The entire build process takes place on a polished perforated plate which makes defined reference points and surfaces from the CAD reproducible by using special clamping and positioning devices. It also serves as a platform for the subsequent classical finishing treatments. Design details include a build chamber sized 250 · 250 · 185 mm enclosed on all sides, an adjustable build platform inside the chamber, a reservoir on the same level beneath the platform, an overflow container on the opposite side, and a one-directional coating system that travels over the build plane twice per coating process. The coating system is designed as an asymmetrical blade. The scanner mirrors are cooled to avoid thermal warpings, thereby increasing accuracy. The machine is optionally fitted with an integral nitrogen generator. The oxygen content is monitored. The machine has an on-line calibration for continuous monitoring and readjustment of the laser beam during the build process.

Figure 3-38 shows the basic model, the M250, and its entire components.

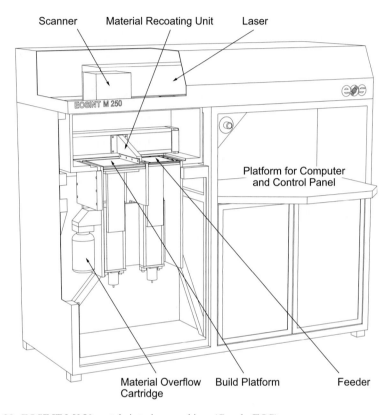

Figure 3-38 EOSINT M250 metal sintering machine. (Graph: EOS)

EOSINT S – Sand Sintering Machine

The development of complex castings, that is, cylinder heads, can be extremely accelerated if complex cores are derived directly from the CAD and produced by the sintering process. Based on the EOSINT concept, a special machine known as EOSINT S700 was developed for the direct sintering of polymeric-coated casting sands. The development was aimed at producing a larger build chamber that matched the requirements of engine design and increasing the process speed in line with the size of the models. These additional requirements were to be compensated, at least partly, by omitting all unnecessary machine properties with regard to the sintering of casting sands.

The sintering of polymeric-coated casting sands is not done by following the principle of liquid sintering, but by polycondensation (see also Section 2.3.1). The machine possesses a build chamber of 720 · 380 · 400 mm and can produce the cavities and cores of typical engine components such as crankshafts or cylinder heads in one piece. The construction with two laser scanner units working simultaneously allows a material consumption rate of approx. 1 liter of sand per hour. A feeding system ensures the continuous supply of molding sands (Fig. 3-39).

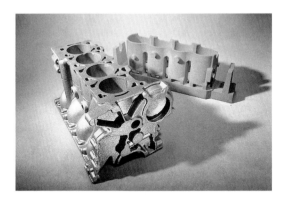

Figure 3-39 Sand casted part (front) and sintered sand core (rear). (Source: EOS)

Material/Build Time/Accuracy:

Material – Plastics

The range of material that can be used in the EOSINT machines includes polyamides (similar to PA12) of various particle sizes with or without glass fillings. Until now polyamide material of an average particle size of 80 to 110 μm was the standard material. This has been replaced today by fine-polyamide with particles of 50 μm. Polystyrene serves as meltable material for precision casting. Polystyrene can be used at relatively low temperatures which reduces shrinkage and improves model accuracy.

Some foundries prefer polystyrene even though in theory it has no real advantages over polycarbonate. Because polystyrene is produced in large quantities (partly because it is also used in a similar configuration for toners in industrial copy machines), it could lead to a potentially drastic reduction of material costs.

When precision casting polystyrene models in a mold the surface must be sealed first to prevent ceramic material seeping in and second to achieve a surface as smooth as possible for the casting process. This is extensive manual work but it is necessary for all (amorphous) sinter materials and thus for prototypers of other manufacturers as well. Using plaster mold castings, which in the course of prototyping have experienced a renaissance, it is not necessary to seal the surface.

Polyamide parts (nylon, PA12) can attain stabilities of up to 90% and in single cases even more compared to injection-molded parts of the same geometric form. They are therefore oustandingly suitable for functional models or for the simulation of injection molded parts.

Material – Metals

The EOSINT M250Xtended is specialized in sintering metal powders for tool inserts by the DirectTool process.

Direct Metal 100 V3 was developed from the EOS-Electrolux bronze powder, has relatively rough granules (approx. 100 µm) and therefore builds faster, but does not generate such good surfaces as the powder with a granules of 50 µm. The component is often infiltrated with epoxy resin especially to improve the surfaces. The tool inserts are best suited for the precision casting of soft or rubber-elastic materials.

Direct Metal 50 V2 is a metal powder of finer granulation but with otherwise similar properties and is therefore especially useful for the production of tool inserts with higher demands on detail and smooth surfaces and for the use of materials that are not too abrasive. Vulcanized molds and blowing molds can be directly used with the achievable surfaces.

For injection molding the cavity is infiltrated with low melting point alloys or preferably today with epoxy resin depending on the geometry and the injection molding compound. As a matter of course the inserts must be manually polished before they are used as injection molds, as is the case with conventionally produced ones also. Experience shows that the material can be processed either by machining or by eroding processes without difficulty [SER95]. The molds may be used for the production of injection molded parts in small series or in preseries of some hundred to several thousand pieces from various plastics such as polyethylene (PE), polypropylene (PP), acrylonitrile butadiene styrene copolymer (ABS), or even glass-filled nylon.

Direct Steel 50 V1 is a powder material on a steel basis. The name already underlines the claim of using pure steel powder in a commercial prototyper for the very first time (12/1998). The information supplied by the producer "without plastic binders," the melting temperature "above 700 °C," and the resistance to extension of approx. 500 N/mm² signal that, by appropriate optimization of the alloy compounds, a relatively low-melting material was developed with properties that resemble those of steel castings. The significant step, however, was to demonstrate how to sinter a dense (density = 7.8 g/cm³) component directly on the prototyper without the necessity of elaborate follow-up processes and thus without the resulting distortion. It is neither necessary nor reasonably possible to infiltrate the component owing to its high density. The material can be machined and wire-eroded.

The surface quality can be improved to approx. $R_Z = 20$ µm by a process called microblasting. Surface qualities up to $R_Z = 3$ µm can be achieved by the polishing methods usually used in tool manufacturing.

DirectSteel 20 is a 20 µm powder which improves surface quality and precision of the parts by allowing a layer thickness of 20 µm.

The perforated plate used for the DMLS process can be fixed directly onto machine tools.

Material – Sands

The range of material includes quartz sands and zirconia sands for applications at higher temperatures. They can be used directly in sand castings and deliver series-identical prototypes.

Build times and accuracy typical for laser sintering can be achieved (see also Section 3.3.2.1).

Post-Processing: For the plastic sinter models the post-processing is identical to the post-processing typical for laser sintering as described in Section 3.3.2.1.

The metal model is submerged into the nearly cold powder bed after the sinter process is finished so that only the powder needs to be removed into the powder container after the build chamber is raised. The build platform can then be easily unscrewed from the elevator. Supports must be removed manually, depending on the material. The platform needs to be smoothed afterwards and should be completely polished after several build processes.

Process-Typical Follow-Up: All finishing techniques and casting processes typical for laser sintering are possible (see Section 3.3.2.1).

3.3.2.4 Selective Laser Melting – Fockele & Schwarze

FS-Realizer^{SLM}, identical with MCP-Realizer^{SLM}

F&S Stereolithographietechnik GmbH, Paderborn, Germany;
Distributed by: MCP-Group (mcp-group.de)

Short Description: Prototyper for the sintering of metal- and ceramic powders. The machine works on the laser scanner principle. A IR laser (Nd:YAG) scans the surface of a powder bed and incipiently melts the particles that form the models after solidification. The process requires no additional supports. The process runs with almost any type of metal or ceramic powder, such as stainless steel, titanium, silicon carbide and aluminum oxide.

Range of Application:

Metals: technical prototypes, direct tooling (series parts by means of tooling), functional metal parts

Ceramics: molds made from ceramics

Development State: Commercialized since 2002 distributed by the MCP-group.

History: see Section 3.3.1.4.

Development Partners/Strategies: F&S and the Fraunhofer Instutute for Laser Technology (FhG-ILT,Aachen, Germany) formed a close partnership for the development of this system. One of the strategies is to be independent from proprietary powders by using only one component powders from the market.

The prototyper will be distributed by the MCP-group who traditionally deals with alloys, fine chemicals and high purity metals and who is entering the Rapid Tooling market by selling materials and equipment for short run tooling. The will help the small company to ship a sufficient number of systems and to provide a appropriate service.

Data Formats/Software: The machine is operated by a Windows 98 or Windows NT surface. The special software compensates shrinkage, controls the beam width compensation and provides all kinds of Rapid Prototyping functionality's such as data repair, slicing, hatching and part positioning on the build platform.

The Principle of Layer Generation: The laser scanner systems principally works as pointed out in Section 3.3.2.1. Because of the development partnership with the FhG-ILT the entire process is very much alike the SLPR process (see Section 7.2.2).

System Type/Construction: The system is operated by a 100 W solid state laser scanner device. The dimension of the building chamber is 250 · 250 · 240 mm (9.85 · 9.85 · 8.1 in.) The machine has a powder storage of approx. 0.53 cu. ft. which e.g. means 265 lbs. of steel powder. It is equipped with a shielding gas device that delivers approx. 0.1 cu. in of inert gas.

Material/Build Time/Accuracy: The system produces fully dense parts. The building speed is approx. 0.3 cu. in. per hour. This definition shows, that the recoating time must be quite short and the build time, therefore, independent from geometry.

Post-Processing: No process related post processing such as infiltrating or sintering is required. The part can be handled as any part made from the same material - steel parts can be machined by any kind of machine tool including EDM. It can be polished to mirror quality.

Figure 3-40   FS-Realizer[SLM] (identical with MCP-Realizer[SLM]), exterior view. (Source: F&S Stereo-lithographietechnik)

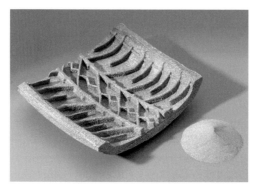

Figure 3-41 Mold for a tire, SLM process (Source: F&S Stereolithographietechnik)

Figure 3-42 Tool insert with conformal cooling channel build by the SLM process: Principle, cut
away picture, part. (Source: F&S Stereolithographietechnik)

3.3.2.5 Laser Cusing, Concept Laser

M3 linear

Concept Laser GmbH, Lichtenfels, Germany

Short Description: Prototyper with interchangeable modules for the sintering of metals,
laser engraving and laser marking. The machine works on a combined gantry and laser
scanner principle. A IR laser (Nd:YAG) scans the surface of a powder bed and incipiently
melts the particles that form the models after solidification. The process requires no additional supports. The process runs with genuine steel powder. The parts do not show
shrinkage or distortion.

The modules for engraving and caving as well as for marking can be used with any kind of
specimen.

Range of Application:

Metals: technical prototypes, direct tooling (series parts by means of tooling), engraving and caving, marking

Plastics: marking

Development State: Introduced in 2002, commercialized since 2002.

History: Concept Laser Ltd. was founded in 2000 as a autonomous subsidiary of the Hoffman group, which is basically a traditional tool maker who also runs a Rapid Prototyping service bureau.

Development Partners/Strategies: The system is a shop floor device not only for making mold and die inserts but for engraving, carving and marking it in the same machine using interchangeable modules. The customer gets three laser processing units by buying just one.

Data Formats/Software: Apart from a solid state Rapid Prototyping software there is a special software that controls even 3D engravings and carvings on the top of solid surfaces.

The Principle of Layer Generation: The laser sintering system principally works as pointed out in Section 3.3.2.1. The entire build process called cusing, a agronomy of cladding and fusing, indicates that it is probably dominated by a liquid phase. It is claimed, that the whole layer is processes at once, hence allowing to avoid distortion of the part.

Although it is very much alike the other RP sintering processes it shows a high speed motion system and a special shielding gas application.

System Type/Construction: The system is operated by a solid state laser in the 100W range. The high speed motion system is based on the combination of scanning devices and gantry type high speed linear drives. It is equipped with a circulating shielding gas device. The laser beam diameter can be adjusted either to increase precision or build speed. The dimensions of the build chamber for cusing are $250 \cdot 250 \cdot 170$ mm ($9.85 \cdot 9.85 \cdot 6.7$ in.). For engraving or carving and marking there is a 2D platform of $450 \cdot 450$ mm ($17.7 \cdot 17.7$ in.). Engraving or carving can be done in all three dimensions. The build chamber allows the handling of workpieces of max $800 \cdot 500 \cdot 400$ mm ($31.5 \cdot 39.4 \cdot 15.75$ in.).

The system is equipped with a automated powder extraction device, which allows the easy removal of the part and in combination with the circulating shielding gas, the treatment of the whole surface after processing.

Material/Build Time/Accuracy: The system produces fully dense parts. The build speed can be estimated from various case studies to approx. 0.5 cu. in. per hour. The material properties show the typical effect of laser material treatment: exaggerated yield strength, and decreased elongation.

Post-Processing: After the cusing process a (confidential) surface treatment is done in the machine to achieve high surface quality and hardness.

No further process related post processing such as infiltrating or sintering is required. The part can be handled as any part made from the same material. Steel parts can be machined by any kind of machine tool including EDM. It can be polished to mirror quality.

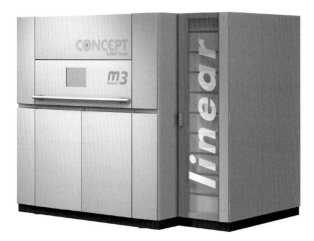

Figure 3-43 M3linear prototyper, exterior view. (Source: Concept Laser GmbH)

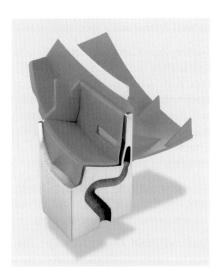

Figure 3-44 Laser cusing process, workpiece with conformal cooling channel, cut awy view (Source: Concept Laser GmbH)

Figure 3-45 Laser cusing process, tool insert with conformal cooling channel, cut away view (Source: Concept Laser GmbH)

3.3.3   Layer Laminate Manufacturing (LLM))

3.3.3.1 Machine-Specific Basic Principles

Layer laminate processes or layer laminate manufacturing (LLM) are often called laminated object modeling (LOM). The expression LOM process has become accepted in the meantime and is used as a synonym for the entire process family[1] even though this is a manufacturer's brand name (registered trademark of Helisys Inc., USA, now out of business).

The Principle of Layer Generation: The LLM process is the direct implementation of the theory of layer processes by a machine (Fig. 3-46). A CO_2 laser cuts out the contour according to the relevant layer information by means of a scanner or a plotter. In the strict sense this is therefore a combined removing (cutting out) and generating process. Either a layer material laminated with polyethylene adhesive is used, or the adhesive is applied in the course of the layer generation. The 3D model is generated in the z-direction by connecting (in practice mostly "gluing") successive layers.

Because of the construction principle, only the outer contours are cut out. The attractiveness of the LLM process, therefore, compared with the raster scanner process, increases as the solid portions of the cross section grow in size. The process is especially advantageous therefore when relatively massive models with large model volumes are to be built. Special hollow structures such as "Quick Cast," "Hülle-Kern" *(Skin and Core)*, "Fast Sculp," and so forth are in principle not possible.

Supports are not needed. Depending on the process, those parts not belonging to the model cross section either have to be removed as chips after each layer is contoured or remain in the model to be removed later. In this case they are cut into small squares (hatched), resulting in small cubic residues within the model. After the model is taken out of the machine the parts must be removed.

The construction method dictates that structures that are internally enclosed can be built hollow only if the supporting internal parts are removed from each layer. The same applies to a certain extend to models that are largely undercut. For this reason deep internal drillings that may be difficult to reach can not easily be cleaned and therefore be represented in great detail.

System Type/Construction: The basic principle was taken up by several producers who developed various process families:

In very simple *(semi-) manual layer processes,* each separate layer is cut out of paper with a cutting knife plotter (JP System 5, Schroff Development Corp.) or out of polystyrene with a cutting plotter equipped with a hot wire. The model contours are afterwards taken out and manually glued together to form the model. These processes are of low cost and allow, owing to the relatively thick polystyrene material, the construction of large but rough

---

[1] See also the remarks at the beginning of Chapter 3

models. It is a disadvantage that the quality of models depends on the manual skill of the user. Both processes are much discussed, but are not, or not yet, available in Europe.

Automatic (paper-) layer processes have a bigger market share. The Helisys LOM machines and the nearly identical prototypers of Kinergy (Zippy) are two examples. They cut out the contour layer by layer with a laser, hatch the material that does not belong to the model but leave the portion that is and generate the whole block with the model inside. When paper is used they deliver wood-like models with a surface that can compete with others. Other materials such as plastics, ceramic composites, and metals are already available or have been announced. The post-processing, that is the removal of unrequired model parts, is time consuming and difficult.

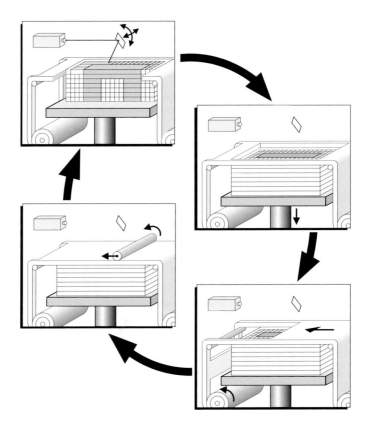

Figure 3-46 LLM process, principle

Relatively new are *layer milling processes* where layers of up to 20 mm are continually contoured by milling in the x-y and especially in the z-direction. Models processed like this do no longer have steps. Two such machines are the Layer Milling Process (LMP; F. Zimmermann, Germany) and the Stratoconcept PC-Process (charlyrobot, France). The

Stratified Object Manufacturing (SOM, ERATZ, Germany) belongs to the same family. This does not necessarily split the model into layers but into arbitrary undercut-free model parts that are milled and then joined together to form the model.

Post-Processing/Follow-Up: LLM models (Fig. 3-48 and 3-49) are basically suitable for all kinds of casting techniques and follow-up processes in which the model is considered to be the master. In principle, precision casting, vacuum casting, as well as sand casting processes are all possible. At times the production of the models can be more difficult because the paper models are strongly hygroscopic and tend to fan out like puff pastry (delaminate) at freestanding corners and edges. Mechanical finishing work and sand blasting therefore should be carried out with the utmost care. The model should be carefully protected against humidity or its surface properly treated. Figure 3-49 shows an example for vaccum casting based on an LLM model.

Precision Casting

The LLM process is basically also suitable for precision casting. The LOM model is then considered a lost mold. Although it could be melted out (burned out), it should be remembered that high amounts of ash are to be expected owing to the solid construction. Depending on the form, utmost care has to be taken to avoid shell cracking.

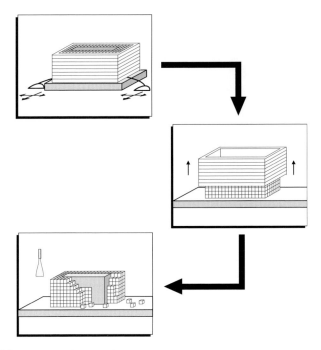

Figure 3-47 LLM process, post-processing

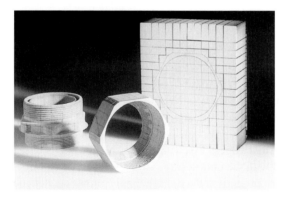

Figure 3-48 LLM model before (rear, right) and after (front, left) post-processing

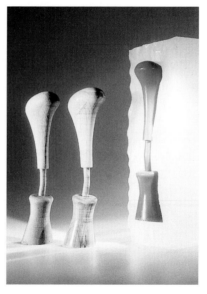

Figure 3-49 LLM model before (left outside) and after (center) finishing. Vacuum casted part (rear, right) with silicon mold

A better choice is the indirect process, in which the LLM model is realized as a separable negative that can be used for the production of wax models. In this way hundreds of wax models can be injected into one LLM mold which are then processed further to casting clusters in the classical precision casting model.

Sand Casting

Sand casting is a typical process for large series and is designed for several thousand castings. LOM models are most suitable when up to approx. 100 castings produced by a series-similar process and with the series material are required. LLM models are best qualified because they produce very stable molding boxes and models. After the build process they must be finished (polished), sealed, and varnished and can be used directly for sand casting.

Owing to their woodlike structure founders accept LLM models as a replacement for their well established wood cores. Casts can be made directly without the foundry having to feel its way into a new technology by time-consuming experiments. LLM components are the ideal supplement for laser sintered sand cores for the fast construction of complex molding boxes.

For larger numbers of parts, LLM models are molded by silicone casting processes in hard plastic material or by precision casting processes in aluminum or steel material. Depending on the complexity of the geometry the LLM model can also be copymilled directly in the steel material.

Thermoforming

Their stability make LOM parts suitable for use directly in vacuum deep-drawing processes (thermoforming) provided they have undergone an appropriate pretreatment. They resist temperatures that occur during these processes so that polycarbonate, ABS, propylene, and similar plastics can be used. It is also an advantage that the models can be made very large and that double layer thicknesses can be used for large models. Furthermore, the production tools can also be generated on the basis of these prototypes in combination with sand casting processes.

After having been finished thoroughly LLM models may be used for vacuum casting.

3.3.3.2 Laminated Object Manufacturing (LOM) – Cubic Technologies / Helisys

LOM 1015, LOM 2030H

Cubic Technologies, Carson, CA, USA

Short Description: The model is build out of foil material, preferrable out of paper, which is coated on one side with a thermo-activatable adhesive material. Layer by layer is heated and glued simultaneously onto the build platform, or onto the partly finished model respectively. By means of a CO_2 laser the contour of the component is cut out. All parts which do not belong to the entire model are cut into squares and stay in the model block. After the model is finished the block is taken out of the machine and those parts not belonging to the model are removed. The process is of greatest advantage when large massive models with greatly varying wall thicknesses are required. In principle all materials can be used if they are available as foils. Plastics and ceramics are available but not yet very common.

Range of Application: Solid Images, Geometrical prototypes, functional prototypes.

Development State: Commercialized since 1991.

History: Cubic Technologies succeeded Helisys in late 2000. Helisys was established at Torrance, California, USA by Michael Faygin 1985, even before 3D Systems and is therefore (apart from Denken, J., 1985) the oldest company providing prototypers. However, marketing did not commence until 1991. The machines have hardly changed in their basic concept and appearance, but they have been improved in many details. The LOM 1015 was followed by the LOM 1015plus and the LOM 2030 was followed by LOM 2030E and the LOM 2030H. The range of material was extended to plastic (LOMPlastic) and ceramic Composite (LOMComposite). It is probably due to their early market success that directly related systems such as Kinergy (1996) of Singapore and basically related systems such as Kira (1994) of Japan are crowding the European market. In the late 1990's business declined and Helisys was defunct in late 2000. The company was taken over by Faygin's now publicly held company Cubic Technologies. As can be seen on the company's website, the prototypers, the material, and even the trademarks are entirely the same.

People speak about a new LOM machine, called LOM3000, which is equipped with a laser scanning cutting device instead of the gantry type one used in today's prototypers. There is not official release until now (3/2003)

Development Partners /Strategies: The American company Helisys was the first to commercialize the LLM process and offers two basically similar machines named LOM of different sizes for the fully automated production of 3D foil models. These machines cover the same market segment with regard to price and quality as stereolithography, laser sintering, and other (full size) rapid prototyping processes and are therefore direct competitors. Recent development research is concentrating on new materials. Development partnerships exist with the University of Dayton / Ohio on the field of LOM models with curved surfaces.

Data Formats/Software: The only input-data format is the STL format. The software for the LOM installation is equipped with reduced CAD functionalities and enables preparatory operations that are graphically supported on the screen. Data errors cannot be corrected, but the program is very tolerant to errors. In the latest version, the file sizes are no longer limited. The slicing takes place on-line with LOMSlice. It could happen that the machine has to wait for the slice data if the models are very complex. The accuracy of the components in the z-direction is improved as the calculation for the following layer is always based on the current model height. This z-height is measured after each finished layer, albeit at only one point.

The software does not compensate for the beam width. The installation is calibrated by taking the measurements of a defined test piece. The actual values are fed into a program that then automatically calculates certain correction values. The next software will include a capability to compensate for the beam width.

The Principle of Layer Generation:

Mounting the Models and Starting the Process: The orientation of models is more or less as in the SLA. The top side of the layer is usually relatively brown with clear markings of the basic raster, whereas the bottom side is relatively white and hardly shows any raster signs.

Before the build can start some manual operations are necessary. The metal build plate of approx. 10 mm thickness is moved to zero height and presses against the paper from underneath. A sheet of paper is cut out in such a way that it exeeds the planned model size by approx. 20 mm in the x-y direction. The loose internal sheet of paper is removed and the resulting open space on the build plate is covered with double-sided adhesive tape and the sheet of paper is replaced manually. This serves as the basis for starting the build process which now commences. Onto this first paper layer a base is built – a paper block consisting of cuboids 3 to 8 mm high, depending on the model size, and with a base of 25 · 25 mm. This base prevents curling at the start of the build process which could amount to several millimeters and depends on the direction of rolling. The actual build process then starts. Because each of the first few layers of the component needs to be examined, approx. 1 to 2h is necessary for starting the installation including the first check of the component. During this time the process needs to be checked occasionally. The speed with which the roll of paper is reeled out can be adjusted by software. If large components are being produced and therefore only narrow frames remain around the cutout areas, the paper could rip if the reel-out speed is too high. This happens relatively seldom and will, in practice, be noticed automatically by an idling tactile roll. The machine is then automatically stopped. It is possible to start the machine again immediately after the paper has ripped without loss of quality. If the machine is restarted later, after the paper has cooled down, and has, as a result of environmental conditions, taken up moisture and has expanded, the danger of faults such as a mismatching arises.

Build Process: During the build process each new paper layer is pressed on by a roller. The roller is heated up to 330 °C, depending on the type and thickness of the paper to activate the glue. For large parts, the selected temperature is usually slightly higher than for smaller parts. The roller pressure too must be higher for large parts.

Up to four paper layers can be processed at the same time. The highest accuracy results if only one layer is processed. The height of the steps as well as the required power are then at their lowest. If two layers are processed simultaneously the accuracy is still good. The build speed of decreases, however, and the components become somewhat darker. The simultaneous processing of three layers is also possible and works relatively well, but it is seldom done. If four layers are to be processed, the laser performance has to be raised to such a degree that the paper usually carbonizes heavily (black edges).

The grid spacing, which determines how the parts not belonging to the model are cut up, can be varied by software in the z-direction. For fine structures in the x-y direction squares of approx. 30 mm are chosen. Horizontal areas can be automatically rastered very much finer. Here, a grid spacing of usually approx. 10 mm is chosen. Otherwise it is hardly possible to remove the remaining cubes from the horizontal areas. These cross hatches, however, are implemented automatically by the software only if the areas are absolutely horizontal. With nonhorizontal areas the problem arises that very wide layers between the single steps become glued together. These are then partly removed during the cleaning.

If the build plane is not completely filled by the model a relatively high amount of paper is waste. If the amount of adhesive lies below 10% those laminated remains that are not part of the component can be disposed with the normal household garbage. In consideration of the amounts of polyethylene adhesive involved which are not insignificant and future ecological audits, some rethinking is probably necessary.

According to the type of model therefore a large share of the material used is not required for the model itself but for the surrounding frames and internal structures. Components formed as a housing relegate up to 90% of the build material as waste. The price of paper being so low, and disposal presenting no problems, this is not very important. Because the LOM^{TM} processes are basically able to use any kind of material that a laser is able to cut, and this includes practically all currently available materials, this aspect will gain more importance in the future.

System Type/Construction: Helisys offers two installations of different sizes that are basically identical in their technical design, the LOM 1015plus and the LOM 2030H. The 2030H has a build chamber of approx. 813 · 559 · 508 cm whereas the 1015plus measurements are 381 · 250 · 355 cm (length · width · height). Both machines are compact stand-alone installations into which all necessary elements are integrated. The user must provide an adequately dimensioned exhaust system and also, depending on the laser type, a water cooling system. The build chamber itself is easily accessible owing to a larger cover. Large doors in the housing facilitate access to the paper mechanism. The machines are simple to install, with the exception of air intake and exit which must be installed by the user, and they can in principle be run automatically.

The machine works with a CO_2 laser beam that does the contouring. The laser performance and cutting speed can be adjusted in such a way that up to four layers can be cut at the same time. Although the processing of several layers simultaneously increases the build speed, it decreases the model accuracy and leads especially to carbonized edges as the laser intensity is no longer sufficiently high for a sublimation cut. The paper is fed into the machine in the form of reels and is pulled over the build platform. There it is laminated onto the already contoured model parts by a heated lamination roller. The laser then cuts out the required contour. In addition, each layer is fitted with an outer frame that is uniform for all layers and separates the model from the paper reel.

Frames remain on both sides and between the single layers similar to the borders on a film slide that sufficiently link the paper with the roller to move it further along and, after the contouring of one layer is finished, to position the next layer on the build platform. Figure 3-50 shows a look into the build chamber. The borders of the actual build plane area are clearly visible. The paper has just lifted off the build plane and is being moved along to the left so that a new layer reaches the build chamber. The plotting mechanism consists of a gantry movable in the x-direction (paper moving direction) and the laser head movable in the y-direction (paper crosswise direction) which can be seen at the upper right corner of the build area (Fig. 3-50). The laser beam is guided from above right by means of an adjustable mirror mechanism covered by bellows. The paper roll-up/roll-on mechanism lies beneath the actual build platform.

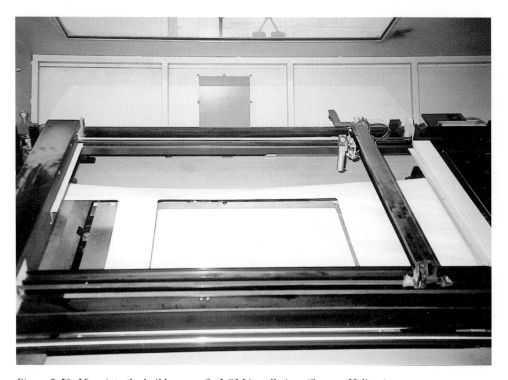

Figure 3-50   View into the build space of a LOM installation. (Source: Helisys)

Model parts that are not needed stay at first in the model. They are in addition cut into small cubes (hatched) because that makes it easier to remove them later on.

Parts that are difficult to remove from the mold can be provided with a "stop" during the construction process. This "stop" makes it possible to partially remove internal, complicated contours from the mold manually and to continue with the build process afterwards.

It is possible to build parts that are movable within one another in one build process provided these parts are already integrated as STL files. The parts are movable because the laser runs over the gapless components a second time, resulting in a gap that makes the parts movable against one another.

The advantage of the LOM process in comparison with other rapid prototyping processes lies in the production of relatively pressure-resistant models that are excellently suited for the simulation of deep drawing, forming, and molding processes for smaller numbers of pieces. The disadvantages of these models are pronounced anisotropic mechanical-techno-logical properties. In layer direction the model is distinctly firmer and more stable than at right angles to the layer direction. This is especially valid for tension load and bending stress. As previously mentioned the compression stress is not much affected.

Material/Build Time/Accuracy: The model is usually made from single layers of thin paper (0.1 mm). Even if the build chamber is operating at full capacity the build progress for complex geometries is sometimes only 2 mm/h. If the grid spacing is very wide and the geometries are simple then the construction progresses at a maximum of 12 mm/h. If the parts are highly filigreed requiring a very fine raster to be scanned (for the remaining material) the components will be much more expensive.

The accuracy is usually ± 0.25 mm for the entire build chamber.

Materials can be polyethylene coated paper foils (LOMPaper), polyester foils (LOMPlastic), and glass fiber reinforced composites (LOMComposite). LOMComposite consists of glass fibers that are orientated in different directions but not interwoven. The adhesive layer consists of thermally active epoxy, while the upper cover layer is acrylate. LOMComposite is 10 times stronger than paper, is moisture resistant, and can resist higher temperatures so that it can be used for functional parts. The LOMPlastic is a polyester foil coated, like the paper, with polyethylene copolymer, an adhesive that can be thermally activated. This can be used without having to modify the machine. The accuracy lie between 0.25 and 0.4 mm depending on the model.

The installation works to an accuracy of approx. 0.25 mm. This could be improved by the manufacturer considerably by using a higher mechanical precision in the plotter control (especially the y-control).

Post-Processing: The metal build platform has holes that provide several possibilities for removing the finished component:

- The adhesive tape is severed by a hot wire.

- The reverse side is warmed up by an iron and the component then removed with a spatula.

- The reverse side of the plate is treated with a solvent that dissolves the glue of the adhesive tape through the holes. The component can then be removed from the plate relatively easily after 2 to 3 min.

It is relatively difficult to remove the waste from the model when it is cold. Therefore, it is either put into an oven directly after production, or it must be cleaned immediately.

First the frame surrounding the component is completely taken off. This may require spatulas and chisels. Then the cubes surrounding and supporting the model are removed and in this way the model is slowly uncovered successively.

When cleaning it should be borne in mind that the laser can cut the layer but not between the layers. The projections of model parts, therefore, protruding in the z-direction upwards and downwards will be stuck to the surrounding structures. This problem increases the flatter the angle of the model becomes and the larger therefore the glued areas become. The cleaning of the model must be done with care (Fig. 3-51).

After the cleaning is finished it is absolutely essential to varnish the models immediately. Otherwise the possibility remains that the components grow especially in the z-direction by up to 1% within 1 to 3 days. The Zapon varnish penetrates the components to a depth of approx. 4 to 5 mm. The components gain a relatively high stability. If they are not varnished their stability remains extremely low.

There are other disadvantages if the components are not varnished. They not only grow due to air humidity but they also tend to become sticky. The adhesive seeping out between the layers in principle prevents the component from being finished with abrasive paper. The edges tend to "fan out" (delaminate).

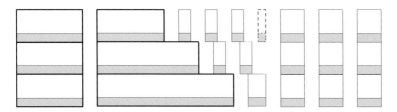

Figure 3-51 LOM process, removing protruding model parts from the mold by additional cuts

The finishing may also be done by milling, turning on a lathe, etc. as long as care is taken with the varnishing and sharp tools are used. The success of milling possibly depends on the direction in which the milling cutter is running. Drilling is possible as well, but care must be taken that the drill does not push the layers apart. Usually this can be done by clamping the components rectangular to the drilling direction.

Even turning the components on a lathe poses no problem as long as absolutely sharp tools are used and an eye is kept on the direction of the lamination. The effort for the finishing work increases tremendously when the component has thin walls. Wall thicknesses from 1 mm upwards can be achieved, but it is not advisable to attempt to build with wall thicknesses of below 1.2 mm because of the effort required for the finishing. To realize components with thin walls the pressure of the roller must be relatively high. Otherwise the components tend to delaminate (separation of the layers) not only after being taken out of the mold but also during the build process.

3.3.3.3 Rapid Prototyping System – Kinergy

ZIPPY I, ZIPPY II

Kinergy PTE Ltd, Singapore

Short Description: The prototypers of the ZIPPY family work on the principle of layer laminate manufacturing (LLM). Their design and function resemble to a large extent that of the Helisys LOM prototypers. The model is generated in layers from paper which is laminated on to the model and then contoured by the laser. In 1999 the ZIPPY II was the rapid prototyping machine with the largest build chamber.

Range of Application: Geometrical prototypes, functional prototypes.

Development State: Commercialized.

Development Partners/Strategies: Kinergy produces mechanically robust prototypers and offers a very large build chamber which was the biggest one in 1999. The hardware and software of the ZIPPY II machine is currently being revised.

Data Formats/Software: The machine reads error free STL files. The read-in takes a relatively long time, sometimes up to 45 min. The slicing is done before the build process starts. The results are stored in temporary slice files. The machine possesses a z-sensor that takes measurements at particular points and defines the height while interpolating the slice files accordingly. In view of the properties of paper that absorbs moisture, and taking the size of the build chamber of the ZIPPY II into account, taking measurements only at particular points is the minimum solution.

The producer supplies only the bare essentials in the way of programs for the operation of the machine with the control computer.

The Principle of Layer Generation: Paper coated with an adhesive that can be thermally activated is transported from a reel over a build platform and glued to it by a heated roller.

A CO_2 laser cuts the contour out of the paper. Areas that are not part of the model are cut into squares so that the model and the waste can be easily separated later. A frame is cut out around the model. The paper is then lifted off, transported along by one model length, and the process starts with the next layer all over again. The paper that is not used is rolled up. The process can be interrupted at any point and then started again. This is important especially when the paper rips. The standstill should not last too long. If the model cools down, it could easily cause a mismatch in the component of several tenths of a millimeter when the system is started again.

System Type/Construction: The ZIPPY II is at present the largest rapit prototyping machine with a build chamber of x, y, z = 1180 · 750 · 550 mm. It is very robust but is also rather slow owing to the large masses to be moved and the plotter mechanism, which is driven by a driving belt. Handling such large "paper blocks" is difficult and requires a well equipped workshop.

The smaller ZIPPY I (x, y, z = 380 · 280 · 340 mm) is faster and less complicated to operate.

Figure 3-52 shows the ZIPPY I and a view into the build chamber with the massive plotter.

Material/Build Time/Accuracy: The paper is available in various thicknesses. Standard is a paper gauge of 0.1 mm. According to the producer a plastic laminate is also available. The paper is relatively cheap, has low warping, and produces little residual matter in the effluent air. Different batches often have various colors so that the unfinished models often appear somewhat chequered has nothing to do with the accuracy, however. The composition of the paper allows the filters to be cleaned with water.

Post-Processing: In these installations the post-processing is identical to the LLM-typical post-processing as described in Section 3.3.3.1.

Process-Typical Follow-Up: All LLM-typical follow-up processes can be used as described in Section 3.3.3.1.

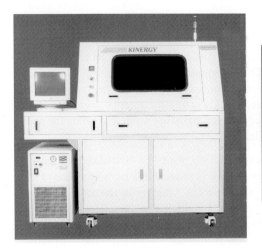 

Figure 3-52 ZIPPY I: Prototyper (left), view into the build chamber (right). (Source:Kinergy)

3.3.3.4 Selective Adhesive and Hot Press Process (SAHP) – Kira

SolidCenter KSC-50N (PLT-A3), PLT-A4

Kira Corp., Japan

Short Description: Paper layer process on the basis of a copy machine. The toner is thermally activated, thereby facilitating the gluing together of the single layers. The contouring is done by a cutting plotter after each layer is stuck onto the model block.

Range of Application: Geometrical prototypes (functional prototypes)

Development State: Commercialized since 1994 in Japan, since 1997 abroad.

More than 150 systems have been sold worldwide (2001).

History: The Kira Corp. was established in 1944 and acts mainly as a machine tool manufacturer. The Rapid Prototyping department was established in 1992. Kira ships to the US but is the the only Japanese supplier of Rapid Protoyping equipment on the European market (2002). The SAHP process is also known as the paper lamination technique (PLT). The SolidCenter PLT-A3 is also known under its original name, SolidCenter KSC-50N.

Development Partners/Strategies: The machines are designed as office printers and consist to a large extent of standard components. Standard paper is used.

Data Formats/Software: Only error free STL data are processed. The number of triangles is currently still limited to 40,000, which is considered a disadvantage.

The RPCAD supplied by the producer enables all operations that are necessary for the production of models to be carried out with error-free STL data and transmits the construction data to the control software. All necessary model and machine parameters are defined by the installation itself. Should the STL-data not be error-free independent rapid proto-typing software must be used.

The Principle of Layer Generation: After the STL data are read in, the geometry is sliced in lagers of a height of approx. 0.085 mm. This information is fed successively into a laser copying machine that prints it on paper. The toner consists of a thermally activated artificial resin. The borders of the contours are wetted thickly while the internal areas are wetted thinner. Areas not belonging to the model are covered by a fine mesh that holds the sheets together but is easily removed when the model is cleaned.

The sheet of paper is placed face down on to the base plate (the first sheet) or onto the already partly built model and the whole paper stack is pressed against a hot plate (175 to 185 °C) that activates the adhesive and glues the paper on. Afterwards, the actual contour is trimmed with a knife controlled by a cutting plotter. In addition to the model geometry so-called release cuts are made that facilitate the removal of those parts not belonging to the model from the paper block after the build process is finished.

System Type/Construction: The PLT-A3 installation consists of (from right to left, see Fig. 3-53) a basically common laser printer (photocopier principle), a process computer and transport unit, and the entire build chamber. The single elements are clearly visible with distinct functions and origin. The installation is compact and can be accommodated in a larger office; its dimensions are 2000 · 1400 mm (including the monitor) and it weighs 600 kg. The process computer takes in all input data and supervises the process. The work platform itself is accessible only by a door on the left side, which is rather difficult to reach. It can be removed from the build chamber only vertically, not horizontally, which is probably acceptable for these relatively small models. Material and toner depots are easily accessible and can be topped up during the production process. The plotter can be fitted with several cutting knives so that a tool exchange during the process is avoided. The machine needs only a (heavy) power supply; devices for outgoing and incoming air are not provided.

Figure 3-53 SolidCenter PLT-A3 of KIRA. (Source: KIRA Corp.)

Figure 3-54 SAHP – PLT process: Modell of a rim (Source: KIRA Corp.)

The PLT-A4 is the smaller version and is meant for the processing of A4-papers. It is significantly more compact; all functions are integrated in one casing.

Material/Build Time/Accuracy: Basically, common copying papers 0.08 to 0.09 mm thick and special toner with thermally activated adhesive are used. Additional toners are presently being developed and tested. According to users, the accuracy in the x-y direction are ± 0.1%, but at least ± 0.2 mm, in the z-direction ± 0.2 %, but at least ± 0.1 mm. The construction speed of approx. 2 to 5 mm/h depends to a large extent on the geometry.

Post-Processing: The model that emerges from the machine as a compact paper block must be carefully freed from superfluous material. If the walls are very thin, 1 to 2 mm thick, the danger of delamination is imminent. It has proved very helpful, especially with inaccessible model geometries, to use additional release cuts that are defined by the software during the preparatory phase.

Process-Typical Follow-Up: To improve the model stability and to avoid swelling the models should be infiltrated with epoxy resin and varnished, if necessary, after they are taken out of the block. If a higher surface quality is required the model can be polished.

The models can be molded in a vacuum casting process after careful treatment of their surfaces. Cast iron and lost wax processes are successfully demonstrated.

3.3.3.5 JP Systems 5 – Schroff Development Corp.

JP Systems 5

Schroff Development Corporation, Mission, KS, USA

Short Description: Rapid prototyping processes consisting of the JP5 software for the calculation of layers of constant thickness from volumes and a cutting plotter for the generation of corresponding paper contours. The layers consist of paper with a one-sided coating of adhesive and are manually assembled. The process is extremely simple and very economical.

Range of Application: Solid imaging.

Development State: Commercialized.

Development Partners/Strategies: The developer aims at a simple model building process that is at the same time economical to purchase and to run and is therefore suitable for school training courses. For these reasons, standard components have been used and automatization in model making has been avoided.

Data Formats/Software: The machine reads error-free STL data or geometric information directly from SDC's CAD software, Silver Screen 3D CAD Solid Modeler.

The slicing is done automatically after definition of the paper gauge as is the model-orientated and waste-optimizing positioning on the build platform.

The Principle of Layer Generation: The generative process largely resembles the principle of layer laminate manufacturing and is closely related to the SPARX Process.[1] The system is based on the JP System 5 Design Slicer software which slices the virtual 3D CAD model into layers according to the paper thickness and arranges the contour information on the construction plane. It also defines the division into submodels and the drillings for the guiding of the entire sheet and single contours. It also defines the cutting geometry. This is always done independently from the physical generation of layers.

The layers are contoured on a cutting plotter and manually joined by means of a gauge after the protective foil is removed.

System Type/Construction: The JP System 5 packet offers optionally a Roland Digital PNC900/910 or a Graphtec FC-3100-60 cutting plotter. The x-y plane is accordingly approx. $215 \cdot 278$ mm or $430 \cdot 556$ mm in size. In the z-direction the model is theoretically not limited since the layers are joined manually. Common construction sizes are approx. $150 \cdot 100 \cdot 50$ mm.

In addition, the system includes everything that is needed for making a model (except a PC with monitor). The sheets are stacked coated side up onto assembly sticks on a build platform, the so-called "registration board." In the area of the model cross section the adhesive foil is removed. There the next layer is glued, and so on. In the next work step, the partial models thereby generated on one sheet are joined together by means of the drillings so that they form the required models.

Figure 3-55 shows the cutting plotter and a processed sheet with contour cuttings, severance cuttings and large and small assembly drillings.

Material/Build Time/Accuracy: Papers of thicknesses varying between 0.1 mm and 0.3 mm are used with a layer of adhesive protected by foil on one side. The paper is available on the retail market. The models, which are usually rather small, are generated in a few hours and cost very little. Their accuracy depend very much on the skill of the model maker and on their division into appropriate submodels. If the work is done with precision, accuracy of

---

[1] The Hot-Plot process of the Swedish company SPARX is closely related to the JP System 5 process. It is based on the CAD program PC-DRAFT solid, consists of its own A3 cutting plotter, and an assembly unit. Model materials are polystyrene (styropor) in the form of sheets of approx. 1 mm thickness covered on one side with adhesive, which are cut by a hot wire on the plotter. SPARX obviously went off buisiness. The system is no longer sold.

between ± 0.5 to 1.0 mm in the x-y direction can be achieved. In the z-direction the accuracy decrease with the increasing height of the model.

Post-Processing: The model must be "threaded" together to partial components and then be assembled from these. The model has to be impregnated immediately, preferably by white cold-setting adhesive. Thereafter it is possible to polish and varnish it.

Process-Typical Follow-Up: It is not advisable to use molding processes because they lack accuracy and the reproduction of details is poor. Vacuum casting can be used in principle after thorough finishing.

 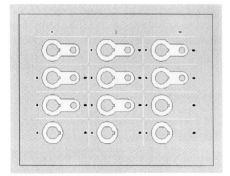

Figure 3-55   JP Systems 5 prototyper of Schroff Development Corp., pattern (Source: Schroff Dev. Corp.)

3.3.3.6 Layer Milling Process (LMP) – Zimmermann

CNC-Schichtfräs-Zentrum LMC (CNC-Layer-Milling Center LMC)

F. Zimmermann GmbH, Denkendorf, Germany

Short Description: Machine (LMC) and process (LMP) for the production of models with large hollow spaces and steep walls. The layer milling center is basically a high-speed model milling machine that contours the model undercut-free layer by layer without stair steppings and under almost constant milling conditions. The center includes a plate storage and the gluing unit.

Materials are model plastics, aluminum, and graphite.

Range of Application: Functional prototypes, technical prototypes, series parts.

Development State: Commercialized since 1998.

History: Zimmermann is well known as a producer of machines for pattern and mold making, but especially as a producer of model milling machines. With layer lamination processes they approach a new line of buisness.

Development Partners/Strategies: The machine was designed to counter the most frequent problems occurring during the production of tools and cavities with large hollow spaces. Such geometries, especially if they are additionally fitted with ribs, require long tools and additional axes both of which have a negative influence on the working time, efficiency, the achievable accuracy, and the surface quality. The LMP process reduces the milling activitey because it always uses the same plate thickness (approx. 20 mm). It works in this way not only with relatively short tools but also under very nearly constant milling conditions. The model, which has been cut into layers mathematically according to the plate thickness used, is continuously contoured layer by layer by milling. A model is generated without steps – otherwise typical for generative processes – and that, therefore, does not require post-processing.

A project partnership for the layer lamination center (LMC) exists with the company Modellbau Pauser, Schwäbisch-Gmünd, Germany.

Data Formats/Software: The machine takes the data directly from the CAD and calculates the (constant) layer thickness as well as the milling path. The entire milling process runs fully automatic.

The Principle of Layer Generation: A plate storage and an adhesive application station are integrated in the machine. Plates of constant thickness are fed into the adhesive station to be coated with adhesive and glued onto the model block. Then, using normal milling strategies, the contour is milled upside down from beneath into the glued-on plate; after the milling, another plate is glued on (Fig. 3-56). In this way an arbitrarily complicated geometry, whose layer proportions are limited only by the size of the machine, is generated overhead. The part has series character provided series material can be used.

System Type /Construction: The most outstanding speciality of this Zimmermann machine is the apparent and also constructionally implemented reversal of the milling principle. In the place where we would expect to find the spindle, the table is fitted, movable in the z-direction. Similarly, where the table should be, we find the spindle which in this case works overhead. Longer, unsupervised milling operations are thereby possible without the danger of an accident caused by the accumulation of chippings. The "overhead" design not only has the advantage that the chippings fall down, out of the work area, but also it facilitates the application of adhesive and the actual gluing process also.

The machine is a high-speed milling machine with automatically interchangeable tools.

Material/Build Time/Accuracy: Basically any millable material that can be glued satisfactorily can be used. Available are model plastics of all kinds, for example, Oreol and aluminum for the production of molds and also positives and graphite for the production of graphite electrodes.

The machine has a working range of x, y, z = 850 · 600 · 750 mm. The plate thicknesses can vary between 5 and 50 mm; the maximum milling speed is said to be 18 m/min. The maximum revolutions of the spindle is 40,000 U/min; a driving power of 2.6 kW is required.

The accuracy is in accordance with those achievable by milling machines and is therefore distinctly higher than this of layer orientated rapid prototyping processes, especially if the large model dimensions are taken into account. The models are completely free from stair stepping as a result of the method, that is, they are contoured continually in the z-direction. Forms with undercuts cannot be realized. Actually such undercuts are not required in those models for which this machine was designed.

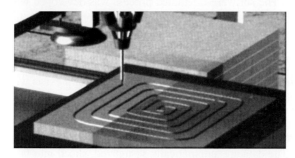

Figure 3-56 Zimmermann layer milling process (LMP): fully automatic plate storage (above), application of glue (middle), upside down layer milling process. (Source: Zimmermann)

Post-Processing: It is not necessary to post-process the model surface.

Process-Typical Follow-Up: The components are preferably used as tools. They are not intended to be used for other moldings although that would be possible.

3.3.3.7 Stratoconception – Charlyrobot

Stratoconception STR 300/.../STR 2000

Charlyrobot, Cernex, France

Short Description: Stratoconception is a software-oriented layer milling technology in which a CAD model is cut into layers so uniform that these can then be contoured by a milling machine and mounted onto guide rods by means of milled-in drilling holes. This method is very much related to the Zimmermann layer milling technology but initially generates single layers that were contoured independently. These can be turned over while being assembled so that arbitrary undercuts and contours can be realized such as those that are required, for example, for the reproduction of human busts.

Range of Application: Functional prototypes, technical prototypes.

Development State: The software-orientated process was developed with the assistance of the French firm Charlyrobot into complete installation for the production of layer milling models suitable for industrial use.

The system was commercialized in 1998.

Development Partners/Strategies: A development partnership exists with CIRTES (Centre d'Ingénierie de Recherche et de Transfer de l'ESSTIN à Saint-Dié-Des-Vosges / University of Nancy, France).

Data Formats/Software: The system reads complete CAD files into a program known as Stratoconcept-PC. The data are automatically visualized, sliced into layers, and assembled into groups that can then be treated en bloc by the milling machine. Afterwards, the fully automatic control program is defined. The machine is also able to use STL files and is therefore part of the "prototyping world", unlike other layer milling machines. The STL file matching the CAD model is read into the program system, visualized, and oriented. The optimal layer thickness for this production is then established.

Principle of Layer Generation: The layers are generated by a continuous milling of the x-y contour in the z-direction.

The model is cut into appropriate layers and these layers are waste-optimized and arranged into partial models and allocated to the available working space. In addition to entire geometry, drillings are made with which the model can later be mounted and positioned with the help of guide rods. Undercuts are realized by turning the model layers for the milling; after completion, before assembly, they are turned back again. Models that have two opposite undercuts in one layer can either not be produced by this method, or they must be divided.

System Type/Construction: The machine itself consists of the PC-controlled calculation station and a high-speed micromilling installation. Depending on the model size, it is available in various construction sizes of x, y, z = 300 · 210 · 100 to x, y, z = 2000 · 3000 · 150 mm in six gradings (300, 400, 500, 1000, 1500, 2000) (Fig. 3-57).

The milling programs are prepared in such a way that the single slices are joined with the frame by very thin strands of waste material enabling them to be taken out of the machine as one unit.

Figure 3-57 Stratoconception prototyper. (Source: Charlyrobot)

Apart from all types of cavities, models with undercuts can be produced as well on the basis of the principle of this method. The process is therefore well suited for the production of busts or similar models and also for the production of relatively large functional parts.

Material/Build Time/Accuracy: Any millable material can be used; the accuracy are reported to be 0.5 to 0.1 mm in the x-y direction. Plastic, wood, and composite material can be processed on machines up to size STR400. Aluminum and other metal alloys can be processed on the higher powered range of machines beginning with the STR500.

Post-Processing: After the milling process is finished the contoured layers must be taken apart and, in a rather intensive manual process, threaded and simultaneously glued and pressed. There is a risk of creating a fault in the z-direction which, depending on the material, would result in the appearance of anisotropic material properties in a direction rectangular to the layer. Assembling the model outside the machine, however, has the advantage that models can be built that are considerably larger than the actual machine build chamber. Owing to the microstrands that hold the milled parts, the model surface has to be finished manually afterwards.

Process-Typical Follow-Up: Molding processes are not intended but possible.

3.3.3.8 Stratified Object Manufacturing (SOM) – ERATZ/MEC

Software-Tool SOM

ERATZ, Dortmund, Germany

Short Description: Software-tool independent of special milling machines for separating complex milling processes into undercut-free suboperations.

Range of Application: Functional prototypes, technical prototypes.

Development State: On the threshold of commercialization.

History: SOM Technology was developed by MEC GmbH, Aachen-Alsdorf, Germany, and Klink GmbH in 1994. After Klink had to pull out of the project for financial reasons little was heard of SOM for some time. The process reemerged at the end of 1998 with new partners and with clear software preferences [NAC98].

Development Partners/Strategies: A development partnership exists with the FH Düsseldorf (University of Applied Science, Düsseldorf, Germany).

Data Formats/Software: Import of CAD data via neutral interfaces (IGES, VDAFS). The development of STEP-compatible interfaces had not been finished yet.

Extensive software for the generation of partial geometries and of the necessary milling parameters is supplied by the producer. It supports visualizing; simple functionalities such as positioning, mirroring, deletions, the definition of milling geometries, and reference areas/positions; and the interactive generation of partial areas. The data export is to a large extent machine independent.

The Principle of Layer Generation: The SOM process is generally based on two process steps (Fig. 3-58):

- The complex 3D part is mathematically split into platelike undercut-free partial bodies.

- A step-by-step machining of the 3D prototype by surface milling the joint faces, contouring the required construction sections, joining the single plates to model parts, and, finally, after several repetitions of these steps, assembling the complete model.

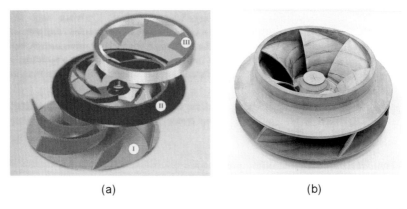

(a)   (b)

Figure 3-58 SOM-process. (a, left) Splitting the geometries into undercut-free parts and production steps. (b, right) Flywheel after assembly. (Source: FH Düsseldorf)

The process steps shown in the example (Fig. 3-58a) are listed in Table 3-1.

Table 3-1 Process steps for the SOM process (Fig. 3-58a)

Step no.	Plate no.	Work sequence	Clamping direction
1	I	first milling step	0°
2	I	first milling step	30°
3	II	first milling step	0°
4	II	first milling step	45°
5	I+II	joining	–
6	II	second milling step	0°
7	III	first milling step	0°
8	II+III	joining	–
9	III	second milling step	0°
10	III	second milling step	20°
11	I	second milling step	0°

This production process solves the support and clamping problems in the machining process and the tool feed is minimized.

It is a disadvantage, in practice at least, that this process is restricted to geometries that can be divided. The process generates a functional prototype but, owing to this production method, the prototype is divided where the subsequent series part is solid. As a consequence, the flow of forces, the stress distribution, and thereby the mechanical stability will differ from those of the series parts in single cases. The result of stability tests, that is, the establishment of the transmission efficiency of the running wheel flywheel shown, will not apply without reservation to the series part.

This process is, however, an interesting and real alternative to the classical rapid prototyping models which are mechanically and thermally significantly less stable than SOM models.

System Type/Construction: The type of systems depends on the milling machine used.

Material/Build Time/Accuracy: The required parts are produced directly from the desired material. All materials that can be machined are usable. Accuracy and surface quality are the hallmark of CNC machines. In addition, CAD integration facilitates consistency in construction data which is typical for rapid prototyping.

Post-Processing: After the milling operation the parts have to be joined. The joining process is not usually included in the SOM process and should be planned beforehand.

Process-Typical Follow-Up: Follow-up processes are not necessary.

3.3.4   Extrusion Processes

3.3.4.1 Fused Deposition Modeling (FDM) – Stratasys

Prodigy Plus / FDM 3000 / FDM Maxum /Titan

Stratasys Inc., Eden Prairie, Minnesota, USA

Short Description: Fused deposition modeling (FDM) means "modeling by application of melted material." The Stratasys FDM prototyper melts bulk or wire-shaped plastic material in a heated nozzle and applies the pasty, melted matter on the model. The layer generation results from solidification due to heat conduction. The advantages of this process are not only the vast range of material including ABS and the ability to use various colors but also the fact that the material used can be changed during the process. The process requires supports. The machines are all suitable for use in offices. Figure 3-59 shows clearly the construction principle.

Range of Application: Geometrical prototypes, functional prototypes, technical prototypes.

Series-identical parts by precision casting.

Tooling by casting processes.

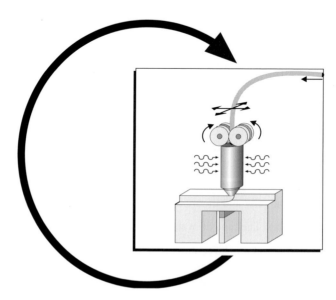

Figure 3-59 Fused deposition modeling (FDM) process, principle

Development State: Commercialized since 1991.

History: Stratasys was established in 1988 and is therefore one of the oldest producers of prototypers. The FDM 1000/1500/1600 prototypers were systematically developed further with larger build chambers and the processes accelerated (FDM 1650/2000/8000). With its dimensions of $600 \cdot 500 \cdot 600$ mm, the FDM Quantum had the largest build chamber of all plastic processing prototypers (status as of July 1999). In 2001 the Quantum was upgraded to the Maxum. The actual program mainly shows machine names instead of type numbers. Besides the FDM 3000 today (12/2002) the family consists of the Prodigy Plus, the Titan and the Maxum.

Development Partners/Strategies: Stratasys is focussing on the competitive process laser sintering and has consequently designed their machines to this end with respect to the size of the build chamber, the series-similar material used, and the achievable speed. This will enable them to produce large, directly usable, functional prototypes of series-similar (nominally series-identical) plastic materials.

Data Formats/Front End: The machine processes complete and error-free STL data.

The producer supplies the *QuickSlice* software that orientes the model in the build chamber, scales or divides it if necessary, calculates the layers and the extrusion path, and generates the construction data. The integrated support generator SupportWorks calculates the required supports automatically. The program transmits the construction data internally to the machine control. This is done with the .sml- (Stratasys Machine Language) data format for the FDM 3000 (and the older 1650/2000/8000 as well) and with the .ssl- (Stratasys System Language) format for the FDM Maxum and Titan. The .sml-file uses an ASCII format and is therefore readable but significantly larger than the corresponding STL file. SLC data can also be read.

QuickSlice is the central processing program (Workstation or Windows NT). The processing of files is done in five steps:

1. Orientation of the parts. The components can, similar to the View program, be scaled or manipulated in the x-y-z-direction by rotating or shifting. As a large part of the construction process is taken up by the generation of supports, the components are preferably oriented in such a way that a minimum effort is needed for the supports (low support volume).

2. Slicing. During the slice process only the outer structure of each component is calculated.

3. Calculation of the geometry (taking the row width into account). In this step the single rows are calculated under a defined angle.

4. Calculation of the supports (SupportWorks). The supports themselves, and each island formed within the component or inside the support, is considered a single element – a so-called set. The supports are generated depending on the calculated row width. The supports have to be placed so close to one another that the distance between them is at the very most half as wide as the row width. If it is too close, the supports are difficult to remove; if it is too wide, the surface of the component will be of very poor quality because the upper layers of the component will be extruded between the supports.

5. Calculation of the base. Definition of the first five layers lying under a bisected angle to the component. Depending on the model material used, the material and process parameters (such as shrinkage factors, speeds, wall thickness, etc.) are supplied by corresponding files and after additional machine specific parameters have been fed in, the control file for the build process is generated.

The Principle of Layer Generation: Thermo plastics and waxes supplied as wire on reels are partially melted in a heated nozzle head and then extruded (see also Fig. 3-59). The nozzle head is controlled by an x-y plotter. After one layer is built the base plate with the model is lowered by one layer thickness and the process starts again with the next layer.

The thermoplastic material (diameter of the hard wire approx. 0.18 mm) is melted in an electrically heated nozzle head (the machine works without a laser) at relatively low, locally limited temperatures of between 68 °C (precision casting) and 270 °C (ABS-plastic). This method has a number of advantages which make it possible to use these machines in an office environment.

The raw material is heated up to just below melting temperature, meets the preceding layer (or the ground plate if it is the first layer) as a semiliquid thermoplastic material, cools off after making contact due to heat conduction, and solidifies. The coating adherence is ensured because the liquified plastic in the nozzle melts the preceding layer locally and temporarily. The distance between the preceding layer and the extruder head and the volume flow of the semiliquid material is adjusted to one another in such a way that track widths between 0.254 and 2.54 mm and layer thicknesses between 0.05 and 0.762 mm result. The distance of the nozzle head to the preceding layer should be about half the nozzle diameter. A cross section of the liquid appears almost elliptical and is named the row width (RW). This extruded billet must always be wider than or as wide as the layer thickness. The optimal layer cross sections have a width/height ratio of between 3.5 and 6. This squeezing of the layer results in solid structures and relatively smooth surfaces, depending on the surface tension and the viscosity (as a function of temperature), and piled up, sausagelike structures are largely avoided.

Although the material solidifies very quickly, supports are necessary for protruding model parts and as a base. The fully automatically generated supports are made of a more brittle material than that used in the model. Therefore, they can be removed speedily, without damaging the model and without tools by manually breaking them off [Break Away Support System (BASS)].

Models 1650 to 8000 have a porous foam plastic plate of about 3 cm thickness as the build platform. The plate is secured in the machine by a frame that prevents its displacement (Fig. 3-60).

Since the porous foam plastic plate is never quite flat the construction process starts somewhat lower than the plate surface. In actual fact, the first layer of the support construction is extruded approx. 0.5 to 1.0 mm beneath the plate surface, that is, the nozzle head runs a groove into the plate surface.

Figure 3-60 FDM 1650 / 2000 / 8000. Base plate of foam plastic for processing plastic materials,
supports, model structure. (Source: Gebhardt)

To stabilize the build process, about five layers are generated at the beginning as supports. The material for the supports is different from that of the actual build material. If ABS is used as model material the support material is a plastic material similar to ABS; for nylon and wax, it is similar to wax. The entire component must be underpinned by supports. For the generation of the supports, the spaces between the lines are half as wide as the track width of the components subsequently created. The angles are defined clockwise from the x-axis. The zero-point of the installation is at the front left hand side of the platform corner.

The FDM Maxum and Titan are building directly onto a thin acrylic plate that is fixed to the perforated base plate by vacuum and is removed with the model after the build process is finished (Fig. 3-61). The situation in the work space of the Titan is quite the same.

The nozzle head usually moves at an angle of 45° to the components in the main direction which is turned by ± 90° after each layer. As the angle for the generation of the supports is only half as wide it is usually around 22.5°. All angles can be altered by the software.

The starting point for each layer can be shifted automatically by the software. With ABS this is noticeable, and if the starting point stays the same, it builds up visibly. It can be polished off relatively easily, however. The construction parameters are adjustable in the z-direction as well as over single areas of any component.

In the course of the build process a radius appears on the outer contours of the corners of the components. This radius forms automatically. In practice the components must always be circumscribed by a border line. The border line is composed of circular elements that for geometrical reasons result in a radius on the outside whereas the outlines of the contours on the inside are always automatically rectangular (Fig. 3-62).

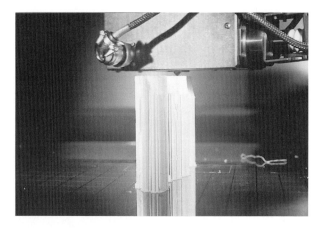

Figure 3-61 FDM Quantum, view into the build chamber. (Source: Gebhardt)

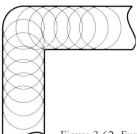

Figure 3-62 Fused deposition Modeling: illustration of the outer and inner corners

A problem is posed by those areas in which the geometry is very filigree. Because the border is moved by the width of half a row toward the inside of the component, similar to the beam compensation in stereolithography, there are often areas in which the border cannot be completely drawn (Fig. 3-63). Since the vectors are also calculated only within the border lines thin structures might not be reproduced. This can be done manually by breaking the set (island) open and calculating the new parameters for this area.

A marginal case would be a border line that ends outside the component and is only one track wide. The ideal line in the middle would be passed only twice, once on the way there and once on the way back, which would result in a rough surface (Fig. 3-63b). The construction process is not interrupted in this case.

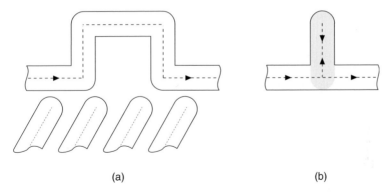

(a)   (b)

Figure 3-63 Reproduction of filigree geometries (a) and thin walls (b)

Figure 3-64 shows a clamp built with the FDM process (right) and half a model with support structures and parts of the base plate (left). Figure 3-65 shows a FDM functional model made of ABS.

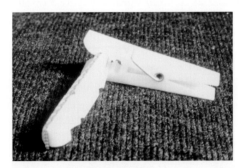

Figure 3-64 FDM model of a clamp (right) and partial model with supports (left). (Source: TNO, NL)

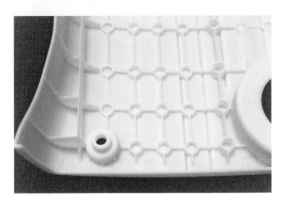

Figure 3-65 FDM Quantum, ABS component, approx. 380 · 600 mm. (Source: AlphaCAM, D)

System Type /Construction: All Stratasys machines have the same characteristics:

- The machines are compact and designed as "plug and play" prototypers.

- They have low power consumption because they have no laser.

- They do not need cooling water.

- For the operators, there are no irritating or toxic vapors requiring extraction.

- They can therefore be run in an office surrounding.

The *FDM 3000* is the standard machine of a family to which the FDM 1650, FDM 2000 and the FDM 8000 belonged. It has, like all the others, two nozzle heads and is therefore able to process build and support material at the same time. No parting compound is necessary between the model and the supports because they are no longer made of the same material. All available materials can be used in the build chamber, which has a size of 254 · 254 · 406 mm. The water works called wash away support system is applicable.

The machine itself is rather compact with a weight of 160 kg. It does not require any special installations. An electricity supply of 230 V/10 A is sufficient. The workstation is connected via a V24 interface. The nozzles are exchangeable. Figure 3-66 shows an FDM 3000 prototyper.

Figure 3-66 FDM 3000 prototyper. (Source: Stratasys)

The *FDM Titan* has a significantly larger build chamber of 355 · 406 · 406 mm and is distinctly larger and heavier but also more modern in its external appearance (but that is a matter of taste). It is designed for the use of ABS, Polycarbonate and Polyphenylsulfone

The *FDM Maxum* (Fig. 3-67) is the largest (build chamber = 600 · 500 · 600 mm) and fastest FDM-machine for plastic material. It is desinged to process ABS.

Figure 3-67 FDM Maxum (Source: Stratasys)

The reason for its speed is the electromagnetic, air-bearing linear drive known as MagnaDrive. It works like a cut open electric drive where the stator is in the base plate and the rotor in the extrusion head. The head is positioned in the x- and y-direction by a moving electromagnetic field. This is done quickly and exactly and, because of the air bearing, very nearly friction-free. Both nozzle heads work on the same base plate but completely independently of one another.

Material/Build Time/Accuracy: Different kinds of materials are available:

- ABS (P400)

- ABS Si (P500), an ABS that can be sterilized for medical use

- Polyamide

- Polyphenylsulfone

- Elastomer (E20)

- ICW – Investment Casting Wax (ICW06), precision casting wax

The material is available in different colors. Therefore colored monochromatic models can be build.

Each material needs its own nozzle head. The nozzle heads have different driving rollers, a different set of gears, and probably different heating elements also, but above all, they have different nozzle diameters. If the material or the nozzle head is changed a test part must be made before the actual build process is started. A cuboid with a base size of 2 · 2 cm and a height of approx. 4 mm serves as the test part. Of the square, only the outer walls are built (borders). On the strength of the connections between the supports and the component it is possible to calibrate the nozzle heads with one another. The construction of this component and its evaluation takes about 15 min.

ABS, the standard material, is a white, relatively impact resistant plastic that has similar mechanical properties to those of nylon. The heating nozzle for the support operates at 265 °C, the heating nozzle for the component material at 270 °C. The envelope temperature is kept between 50 and 80 °C. If the temperatures are too low the layers delaminate, if it is too high the material will be too liquid for the construction process, tend to oxidize, and take on a brown color.

During the build process with ABS the nozzle heads still have to be purged after each layer is finished. This is necessary because ABS tends to create water bubbles inside the nozzles (hygroscopic) which would otherwise result in faults.

ABS Si is a methyl methacrylate ABS for medical use, which can be sterilized with gamma-rays (up to 5 MRad).

The *precision casting wax (ICW 06)* is similar to industrial precision casting waxes and can therefore be used for all known precision casting process without any problem. It fulfills the most important criteria of a low ash content (0.0075%) and a low coefficient of thermal expansion so that shell cracking is safely avoided.

Accuracy of ± 0.127 mm (± 0.05 in.) can be achieved up to 127 mm (5 in.). Accuracy on models greater than 127 mm (5 in.) is ± 0,0015 in./in. (± 0.0015 mm/mm).

Post-Processing: A workshop is not necessary for the removal of the supports. The contact points where the supports meet the model should be polished. Since 1999 a "water works" called support material is available that can be simply washed out.

Process-Typical Follow-Up: All currently known follow-up techniques can be used in combination with the FDM process provided the prototype materials from the available range are balanced appropriately. In single cases, however, it should be checked that the achievable degree of detail, the achievable accuracy, and the resulting surfaces conform to the requirements.

Wax models are designed for the classical precision casting process.

3.3.4.2 Multiphase Jet Solidification (MJS) – ITP

RP-Jet 200

ITP – Ingenieurbüro Dr. Theo Pintat GmbH, Bremen, Germany

Short Description: Machine for the production of models, prototypes, and tools using the extrusion process. A vast range of molten single-component and multi-component plastic granules, powders, and pastes can be used. The machine can also be used for user specific materials.

Range of Application: Geometric and functional prototypes (technical prototypes).

Development tool for materials.

Development State: Commercialization announced by the producer.

History: Dr. Theo Pintat, who influenced the development of the process and the machine considerably, established the ITP GmbH for commercialization.

Development Partners/Strategies: The multiphase jet solidification (MJS) process was developed by the Fraunhofer-Institut Fertigungstechnik Materialforschung (IFAM), Bremen, Germany (Fraunhofer Institute of Manufacturing Engineering and Material Research). In line with its general aims and objectives, the institute strove to develop a simple machine, designed for a broad range of material, offering a vast range of different applications.

The technical realization was achieved by the development partner and machine supplier ITP.

The producer-supplied installation software generates the machine control file and monitors automatically the construction process. Rapid prototyping functionalities are not offered.

Data Formats/Software: The machine reads the geometric data as an error-free buildable STL file.

The preparation of the geometric data and the STL conversion must be done outside the machine.

The Principle of Layer Generation: Thermoplastic material is melted in a heated nozzle head before being applied to the model structure in layers of 0.05 to 0.6 mm. The nozzle head runs over a build chamber of x, y, z = 200 · 200 · 150 mm by means of a triple-axis positioning unit. The model is created by from solidification due to heat conduction in the model.

System Type /Construction: The machine is compactly built around the build chamber and is easily accessible by a hatch that opens upwards (Fig. 3-68). The controlling system and computer are placed beneath the build chamber. A PC workstation for controlling and monitoring the machine is fitted on the side. From there one cannot look directly into the build chamber.

Figure 3-68 MJS prototyper. (Source: FhG-IFAM, D)

The material is fed into a melting chamber and heated to slightly above melting temperature. Prefabricated materials can also be fed into the melting chamber by specially developed cartridges. The build chamber is completely encapsulated to avoid heat loss and minimize noise level. This also allows the maintenance of the enveloping temperature and atmosphere within certain limits in accordance with the requirements of the material. The diameter of the nozzles can be adjusted to between 0.5 mm and 2.0 mm according to the required material distribution.

Material/Build Time/Accuracy: The machine is suitable for all molten and thermoplastic materials that have a melting point under 200 °C and a viscosity range of between 10 Ps and 200 Ps. This includes in practice all materials that are used in plastic and powder injection casting processes and ceramic and metallic powder binder mixtures as found in reaction injection molding (RIM) and metal injection molding (MIM) processes. The material can be in the form of powder, sticks, or granules and can be fed in directly or by means of the mentioned cartridge system. This facilitates special preprocessing; the inclusion of air bubbles, for example, is avoided.

Using the nozzle diameters and layer thicknesses mentioned in the preceding, deposition rates of 0.1 mm³/s to 25 mm³/s (= 90 cm³/h) can be achieved.

Accuracy of 0.2 mm are claimed, which is equivalent to an accuracy of 0.1 % for an edge length of 200 mm.

Post-Processing: A process-specific post-processing is not necessary.

Process-Typical Follow-Up: The finishing depends on the purpose to which the model is to be put. Owing to the broad range of material, all types of models – especially functional models and tool inserts – are possible. The process results in a relatively rough surface and the finishing will therefore focus on this.

3.3.4.3 3D Plotter – Stratasys

Genisys

Stratasys Inc., Eden Prairie, Minnesota, USA

Short Description: Genisys is a 3D plotter following the FDM principle. The machine, suitable for office use, makes polyester models that can be used for design models; additional post-processing is not necessary. The material makes them so stable that practically all tasks necessary in the early stage of product development can be performed with their assistance.

Range of Application: Solid imaging.

Development State: Commercialized since 1997.

History: See Statasys FDM, Section 3.3.4.1

Development Partners/Strategies: Stratasys was one of the first producers to encourage the use of prototypers at the earliest phase of product development, thereby helping to coin the term "concept modeler" and to make this class of prototypers more commonly known.

Genisys was the first prototyper optimized for this specific use.

Data Formats/Software: The software packet AUTO-Gen reads error-free STL files, orientes and scales the models, generates the layer data and the support structure, and generates the machine control files. AUTO-Gen also controls the fully automatic model generation.

The Principle of Layer Generation: Genisys operates with the extrusion process. In the machine this material is liquified and extruded under pressure as molten, pasty matter on to the model (or onto the first layer on the base plate) through an appropriate precision pipe and nozzle system. The material solidifies as a result of heat conduction.

System Type/Construction: The extrusion head is moved by means of an x-y plotter system, while the z-component is realized by moving the build platform. The prototyper is designed as a desktop model.

The dimensions of the build chamber are x, y, z = 203 mm in each direction. Several models can be built simultaneously (Fig. 3-69).

Material/Build Time/Accuracy: The only available material is polyester. It does not use prefabricated plastic wire but rather a polyester raw material in the form of square tablets that are put into a magazine. It is therefore possible to change the material without difficulty in an office location. The machine has a capacity of up to 10 cartridges with 52 tablets each. The material supply therefore holds a total of approx. 2.3 kg. The model and the support construction are made of the same material.

The accuracy as specified by the producer are ± 0,3 mm, which is surely sufficient and realistic for a design model.

Post-Processing: The joints between model and supports are perforated by the software in such a way that the supports can be broken off easily by hand. Basically, no tools are necessary for post-processing. The models can, however, be improved if the support scabs are trimmed properly.

Process-Typical Follow-Up: In principle, design models do not require molding processes. These models should be used only as functional models if it is made absolutely clear that they are concept models and if corresponding optical and mechanical-technological properties are accepted.

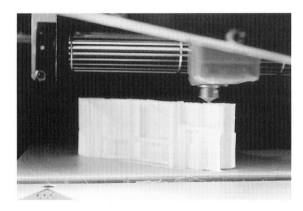

Figure 3-69 Concept modeler Genisys: view into the build space (Source: Gebhardt)

3.3.4.4 ModelMaker – Solidscape / Sanders Prototype, Inc.

Modelmaker II, Patternmaster

Soildscape, Inc., Merrimack, NH, USA

Short Description: Prototyper (Modelmaker II) and prototyping system (Patternmaster) for the production of small, very precise models and prototypes. One print head applies molten hard waxlike material drop by drop while a second applies low-melting wax that acts as support material. Each layer is milled resulting in very precise models that are especially well suited for the precision casting of precision parts. The prototypers are called Modelmaker II and Patternmaker. They are closely related with the Patternmaster optimized for pattern making by the precalibration of five different layer thicknesses.

Range of Application: Rapid prototyping and rapid manufacturing of accurate precision casting components.

Development State: The preceding model, Modelmaker MM6-PRO, has been commercialized since 1994, the Modelmaker II since 1997.

History: Sanders Prototype Inc. was established in 1994 by Royden C. Sanders Jr. The predecessor company, Sanders Design Inc., developed the Modelmaker between 1992 and 1994. Mow it's named Soildscape, Inc., Merrimack, NH, USA

Development Partners/Strategies: The Modelmaker was developed with the aim to facilitate the construction of small accurate components as needed for precision casting processes.

The Modelmaker is not a concept modeler but a functional prototyper for precision parts even if it gives such an impression owing to its appearance and the materials that can be used.

Data Formats/Software: The machine reads the geometric information as STL, SLC, HPGL or DXF files. It operates the propriety ModelWorks Software.

The producer-supplied software ModelWorks reads the data and facilitates visualization, orientation, and positioning. It includes simple CAD functionalities and calculates the supports automatically. ModelWorks defines the construction data and automatically controls the model making.

The Principle of Layer Generation: The heated inkjet print head contains molten material that is shot as microdroplets onto the model structure. The drops then solidify into the model as a result of heat conduction. The impact energy deforms the drop so that the applied material tracks have a height/width ratio of approx. 0.6:1. Layers as thin as 0.013 mm can be realized. Those areas that are not part of the model are filled by a second print head with low melting wax that serves as support material. The surface is milled plano after each layer application so that high accuracy are achieved also in the z-direction. It is not necessary to cool the material.

System Type/Construction: The machine (Fig. 3-70) consists of a build platform over which the two print heads traverse in the x-y direction (position accuracy: 0.006 mm). On the same x-axis a cylindrical milling cutter is attached that is connected to an extraction system. The build platform can be positioned very accurately in the z-direction (0.0032 mm). The useable build chamber is x, y, z = 304.8 · 152.4 · 228.6 mm.

The machine is designed for automatic operation and has material containers with 24-h capacity as well as a device that checks the nozzle function automatically.

The print head emits droplets of molten material with a diameter of approx. 0.0075 mm that are applied in a straight line generating a land of approx. 0.010 mm width and 0.006 mm height. Such a line structure can be generated with a speed of 30.5 mm/s (which is the equivalent of an application of approx. 0.11 mm³/min).

Material/Build Time/Accuracy: The proprietary materials are hard wax blends. The building is done by a green thermoplastic material, called ProtoBuild. The red support material called ProtoSupport consists of natural and synthetic waxes and fatty esters. The building material has a much higher melting point 194°–235° F (90°–113° C) than the support material 129°–169° F (54°–76° C). The material shows no shrinkage, it is non toxic.

The build takes a relatively long time that can be shortened if the layer thickness is increased up to a maximum of 0.13 mm; this, however, impairs the definition.

The accuracy is specified at 0.025 mm over 76 mm, half of each trip (or ¼ of the construction plane) in the x-y plane and with 0.013 mm on the entire z-length of 229 mm (influence of the milling).

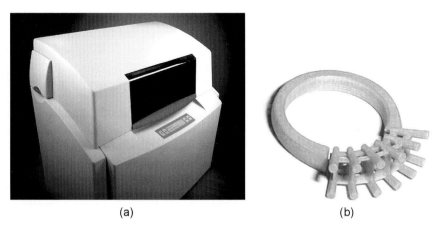

(a)   (b)

Figure 3-70   (a) Modelmaker II, (b) model (Source: SPI)

Post-Processing: The supports are removed with a solvent. Further post-processing is in general not necessary.

Process-Typical Follow-Up: The wax models are especially suitable for precision casting.

3.3.4.5 Multijet Modeling (MJM) – 3D Systems

Solid Object Printer – ThermoJet

3D Systems Inc., Valencia, California, USA

Short Description: ThermoJet is a 3D printer with several piezoelectrical print heads arranged in a line. It was designed for fast and uncomplicated production of design models constructed of a thermoplast similar to hard wax, and as a 3D network printer.

Range of Application: Concept models.

Development State: Commercialized, market introduction in early 1999.

History: 3D Systems introduced the Actua 2100 in 1997. It was a concept modeler, operating on the multijet modeling process. When the Thermojet was introduced production of the Actua 2100 was stopped.

Development Partners/Strategies: The Thermojet is designed as a 3D printer and is supposed to work with a CAD networks in the same uncomplicated way as 2D printers do today. In consequence the producer also describes this machine as a solid-object printer.

Data Formats/Software: After reading the STL data the machine works fully automatically.

The producer-supplied Solid Object Printer software enables the input data to be processed although, according to the producer, this is not necessary for the actual printing process. The user can visualize the CAD file with "preview" and can study it from different perspectives. It can copy the files, scale, turn, mirror, and shift. There is a measuring function that can be switched from inches to millimeters. In addition the software possesses an autocorrection module for incorrect STL data. Several STL files on one platform can be combined to form one image.

The Principle of Layer Generation: Small drops of waxlike thermoplast are applied in layers by means of multiple print heads arranged in a line and functioning on the piezoelectrical principle. The print head has 352 nozzles arranged in the y-direction so that, per layer, the build platform need only be traversed once in the x-direction. The supports are generated at the same time as thin needles of the same material and are easily broken off manually when the model is finished.

System Type/Construction: The ThermoJet prototyper is based on the same base frame as the Actua 2100. It has the dimensions of a larger office phot-copy machine. The build chamber with its x, y, z = 250 · 190 · 200 mm in the y-direction is 10 mm smaller than the 2100. The build progress can be observed through a large window (Fig. 3-71).

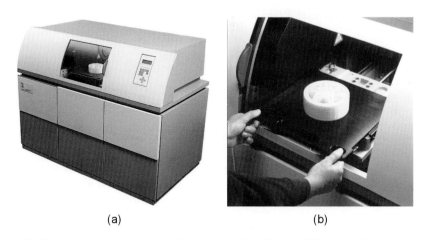

(a)   (b)

Figure 3-71 ThermoJet: (a) prototyper, (b) part on platform (Source: 3D Systems)

According to the producer the reliability of the print heads was very much improved. These print heads currently in use have been tested by processing several tons of material. This aspect is important for the user because the failure of only five nozzles on 100 mm (failure rate approx. 2.55 % in relation to 352 nozzles) causes distinctly visible build faults and obliges the user to change the entire print head. In the case of the Actua 2100 this would cost several thousand dollars.

Material/Build Time/Accuracy: The currently used material is called Thermojet 88 a thermopolymer on paraffin basis is available in white, black, or gray. It is a further development of the thermopolymers Thermojet 45, 65, and 75 (colored Thermojet 65 material) available for the Actua 2100. The new types of material differ from their predecessors mainly in their increased stability with improved ductility so that the danger of breaking the models has been significantly reduced. The number of supports was therefore also reduced so that the models have less "spots" on their undersides than those of the Actua 2100. The material is put into the machine in cartridges of 2.3 kg. (According to the producer, up to 5.9 kg may be put into the machine which is, however, not a whole multiple of this size of cartridges.)

According to the producer the construction speed is three times faster than that of the Actua 2100 which is primarily due to the marked increase in the number of parallel working nozzles from 96 to 352. The maximum definition is 400 DPI in the x-direction and 300 DPI in the y-direction.

Post-Processing: After the build process only the needlelike supports need removing from the model. This can be done by hand.

Process-Typical Follow-Up: Design models on principle require no molding processes. If the models are used for precision castings it should be borne in mind that they were made on a concept modeler.

3.3.5   Three-Dimensional Printing (3DP) Processes

3.3.5.1 Rapid Prototyping System– Z Corporation

Z 402, Z406, Z806 Rapid Prototyping System

Z-Corporation, Somerville, MA, USA

Short Description: A machine family for the production of solid images using the Massachusetts Institute of Technology's (MIT) 3D printing process in which a water-based liquid is injected into starch powder. The process it trimmed for speed and for the simplest of handling. The degree of detail reproduction is low, without a special infiltration the models are not resistant to mechanical stress, and the material is environmentally acceptable.

Range of Application: Solid images and concept models.

Development State: Commercialized as Z402 Concept Modeler since 1998. The Z402C, later called Z406 for coloured models was intruduced in 1999 and upgraded with an advanced nozzle system later. In 2001 the Z804 was introduced.

History: The Z-Corporation was established at the end of 1994 by Marina Hatsopoulos and Walter Bornhorst. The development of a prototyper started after an agreement was signed with MIT concerning the rights to the special use of the 3DP master patent. The Z402 was introduced in 1997.

Development Partners/Strategies: The machine is based on the MIT 3D printing (3DP) patent. It closely follows the concept of a 3D printer. Installation and handling are extremely simple and can be done without prior training. The machine can be run in an office environment, provided powder, vacuum cleaner, and a wax covering and infiltrating station can be tolerated there. Since 1999 making colored models became a strategic goal. With the introduction of the much larger Z804 it appeared that casting applications became a new focus.

A partnership exists with Ciba Geigy (Ciba Geigy Speciality Chemicals) for the development of new materials.

Data Formats/Software: The machine reads the geometric information as an error-free, buildable STL file; the colored models accept VRML (II) and *.ply files .

The producer-supplied installation software generates the slice and machine control files and visualizes and automatically monitors the build process. Other rapid prototyping functionalities are not offered.

The preparation of the geometric data and the STL conversion need to take place outside the machine.

The Principle of Layer Generation: A colored, water-based liquid is injected into a powder bed of cellulose powder by means of an inkjet print head resulting in local solidification and thereby generating the elements of a new layer and joining it with the preceding one. Powder that is not wetted stays in the build chamber and supports the model. The models must be infiltrated with wax or epoxy resin, as otherwise they are not resistant to mechanical stress

System Type/Construction: The machine (Fig. 3-72) consists of a build envelop with three chambers above which a coating and plotter mechanism is installed. Two chambers, the powder supply container and the actual build chamber, have movable bases. The base of the supply cylinder is raised, and a distribution roller takes up a certain amount of powder and, moving across the build chamber, distributes it evenly. Surplus powder passes to the overflow container. The recoating process takes only a few seconds (2 according to the producer).

After recoating the layer is contoured by injecting a colored water-based liquid. The print head works like an inkjet printer with 128 parallel nozzles. The wetted powder particles are firmly linked with one another, forming a physical layer. The unwetted powder stays in the build chamber and supports the model. The liquid accounts for about 10% of the model volume. After the solidification of the layer, the base of the build chamber is lowered by one layer thickness, the base of the supply cylinder is raised accordingly, and the process starts again.

Material/Build Time/Accuracy: The powder is basically starch – chemical terms a polymer of D-glucose. By the injection of a liquid it becomes locally firmly linked. The material can be stored and processed easily, and the disposal of the material is considered to be problem-free[1]. Color is obtained either by one multi color single printheads (CANON type, Z402C) or by six single color printheads (HP type, Z406, Z806)

A large number of powder-binder combinations can be used.

The machine is very fast building, 5 to 10 times faster than other prototypers. Facilitating the speed is a very fast slice process. The coating takes only a few seconds and is also very fast. The print head passes over only those sections of the build chamber that contain a model. Thereby the build time is optimized. Because the print head always passes over the entire area of the above mentioned section, the build time is not influenced by the complexity of the geometry.

The definition in the x-y plane is 600 DPI. There is the additional effect of particles, which really lie outside the contour, being "glued on" by capillary activity and, similar to the sinter process, have the appearance of a "fleece". The models are therefore relatively inaccurate, with a rough surface and a porous structure but suitable for "show and tell" applications (Fig 3-72b).

Post-Processing: The models are taken out of the powder bed after the process and the supporting powder is simply removed by a vacuum cleaner. They can then be carefully removed from the build chamber. Slight sand blasting is of advantage.

Process-Bound Finishing: The models have to be impregnated, preferably by hot liquid wax, especially when they are to be used more often or for a longer period of time. This again influences the accuracy. Alternatively, epoxy resin or an adhesive of the cyanoacrylate type can be infiltrated. Both methods, especially when working with epoxy resin, adds a different perspective to the time advantage and the possibility of using the machine in an office.

Carefull finishing leads to very good results as shown on the example of a tooth (Fig. 3-72c).

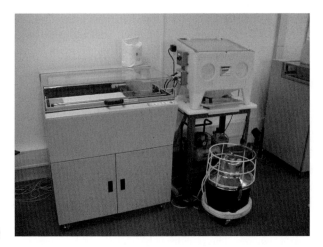

(a)

---

[1] "You have to watch out your dog does not eat the models," Klaus Eßer, Marketing Director Europe, 3D Systems and former Z-Corp / 4D Concepts (Germany) salesman.

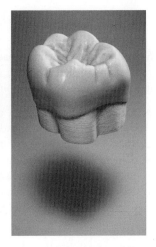

(b)

(c)

Figure 3-72 Z 402 prototyper. (a) Prototyper installation, (b) model, (c) carefully post processed
tooth. (Source: Z-Corp / RP-Lab FH Aachen, Germany; Schmidt / RP-Lab FH Aachen,
Germany)

Process-Typical Follow-Up: Concept models do not in principle require molding processes.
If the models are to be used for precision casting or – after epoxy resin infiltration – as functional models it has to be kept in mind that they are in fact concept models and the corresponding properties must be accepted.

3.3.5.2 Rapid Tooling System – ExtrudeHone

Prometal Rapid Tooling System RTS 300, R4 Rapid Metal Manufacturing System

ExtrudeHone Corporation, Irwin, PA, USA

Short Description: The rapid tooling system 300 (RTS 300) is a prototyper for the production of functional components, molds, and tools made of steel powder using the 3DP
process (license held by MIT). A "green part" is generated in the machine. After the build
process the binder must be removed and the component infiltrated with metal.

Range of Application: Technical prototypes, rapid tooling

Development State: Commercialized since 1997.

History: The development of a rapid tooling system is a new line of business for this
company which specializes in machines and processes for the improvement of surfaces and
finishings.

Development Partners/Strategies: The developers are focusing on the development of functional components and tools of steel, ceramics, and other high melting point materials
suitable for series. The development partnership with MIT and IMAGEWARE combines

with the company's own core competence. The surface treatment of metals is the basis for this strategy.

Data Formats/Software: The machine reads geometric information in STL and SLC formats and has an automatic slice program. The cooperation with IMAGEWARE leads to the assumption that a capable rapid prototyping software exists.

The Principle of Layer Generation: The components are generated layer by layer by injecting liquid binder through a print head onto the surface of the powder bed. In the machine a model, called the green part, is generated. It does not derive its stability from thermal fusion as in the sinter process but by injection of a binder into the metal or ceramic powder. The advantage, in contrast to the sintering of multicomponent metal powders, is that, owing to the separation of the construction material and the binder, a segregation in the powder is ruled out. In addition the process in the machine runs almost "cold."

The binder at first only makes the particles stick together. By exposure to a high-energy lamp the layer is dried and solidified so that a transportable "green part" is generated.

System Type/Construction: The stable machine (weight approx. 3.5t) has a base frame that serves as a portal. It contains the powder supply, the binder supply, and the control unit. The coating mechanism is integrated into the portal. In addition to the print head, a lamp is also fitted to the portal that serves as a heat source mainly for drying the binder. The process chamber lies beneath the portal and has a movable base both in the construction area as well as in the supply area which are moved in the course of the coating by one layer thickness in opposite directions. The process chamber is not sealed off. The maximum size of component is x, y, z = 300 · 300 · 250 mm.

Figure 3-73 shows the machine with its build chamber, coating, and printing installations.

Material/Build Time/Accuracy: Materials are stainless steel powder and ceramic powder. Nothing is known about its specifications or its build time and accuracy.

The process itself runs cold but the drying and the post-processing entail several thermal steps that in general can cause distortions.

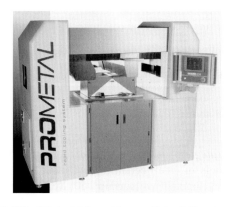

Figure 3-73 Prometal RTS 300 of ExtrudeHone. (Source: ExtrudeHone)

Figure 3-74 Prometal RTS 300: Functional metal part (Source: IFAM Bremen, Germany)

Post-Processing: The green part is solidified outside the machine by thoroughly burning the polymeric binder. Afterwards, the porosities are infiltrated by low-melting (copper-based) metal alloys.

The surface must be reworked mechanically.

Process-Typical Follow-Up: The aim of the process is the production of tools. Follow-up processes are not intended.

3.3.5.3 Direct Shell Production Casting (DSPC) – Soligen

DSPC-1

Soligen Inc., Northridge, California, USA

Short Description: Processes and prototypers based on the 3D printing process producing ceramic molds for precision casting by injecting liquid binder into ceramic powder.

The process, known as direct shell production casting (DSPC) includes the entire process chain starting with the transmission of CAD data up to the ceramic mold ready for baking. Series-identical molds are produced and the process is technically suitable for production (rapid manufacturing).

Range of Application: Technical prototypes, rapid manufacturing.

Development State: Commercialized since 1993, the prototyper itself is no longer sold (1999). The company works as a service bureau.

History: Soligen was one of the first commercial users of the 3D printing technology introduced in 1991. Several of the machines were sold from 1993 onwards. The first industrially usable machine was introduced in May 1994 and was presented on the GIFA fair in

Germany and elsewhere in Europe. Although the producer announced the market introduction for 1995 no machine was sold in Europe. The rapid prototyping process is robust. It seems that so much foundry technical know-how was required, however, that the marketing was terminated and a service with molded parts organized.

General Remarks/Development Partners/Strategies: Soligen adopted the process of 3D printing ($3DP^{TM}$) which had been developed by MIT and developed a machine to produce ceramic molds directly for precision casting processes. This process known as DSPC consists of software and hardware and should therefore be considered as an independent solution. The possibility to produce clusters and therefore several parts simultaneously qualifies this process as a miniseries production process. A general advantage is that the classical multistepped precision casting process, including the model generation, is reduced to two steps – the automatic generation of the shell and the baking. The prototypes are absolutely identical to those of the subsequent series parts in respect to material as well as their mechanical-technological properties.

Data Formats/Software: The machine requires CAD data via neutral interfaces as the basis for the optimization of foundry technology.

The producer-supplied software generates a one-piece mold on the basis of the CAD data for the required mold model, while observing all the technical rules of casting. The basis of the process is a CAD solid of the mold (positive) subsequently produced. The geometry of the ceramic mold (negative) is calculated by the so-called "shell design unit" (SDU), a combination of hardware and software, while observing all accepted casting techniques. These include, for example, scaling to compensate for shrinkage, casting angles, sprues, and vents and monitoring the contour for such details that are awkward to realize in foundry techniques as, for example, breakthroughs or pocket holes. After a geometry is defined, a tree of any size can be built within the restraints of the build chamber with the aid of the software depending on the geometry of the parts. The necessary foundry-technical geometries such as feeders, channels, sprues, and risers are stored in a program library. A special software option allows the casting process including casting simulation and the calculation of metal volumes and weights to be observed and assessed.

The data gathered in this way are read into the machine.

The Principle of Layer Generation: First, the powder feeder applies a thin layer (0.12 to 0.18 mm) of aluminum powder (corundum) onto the build plane. To do this a portal above the build chamber is traversed that helps to distribute the powder evenly over the build plane. With the aid of a roller which moves over the powder bed the powder is evened out and slightly precompressed. The dimensions of the work space are x, y, z = 400 · 400 · 500 mm.

The mold is produced in the DSPC-1 machine in layers. A print head, basically similar to the known design of inkjet printers (speed < 1.6 m/s), injects jellylike silica (collidal silicum compounds, silica gel) into all contour areas that are to have a firm consistency later. The affected powder particles are thereby compounded with one another as well as with the preceding layer. After each layer the piston of the build chamber is lowered by one layer thickness and the next layer is added. The process is schematically shown in Fig. 3-75.

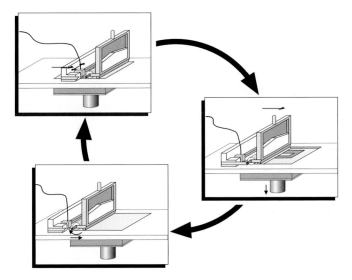

Figure 3-75 Schematic of the DSPC-1 process for the generation of ceramic molds according to the 3DP process, Soligen

The cycle is repeated until the complete mold is finished. It is presintered in the machine and then taken out of the powder bed. The surplus powder is removed with a brush.

In those places where no binder is applied the ceramic particles stay loose and can be removed without difficulty with a brush after the model is finished. The mold is then thoroughly baked, thereby expelling the binder and creating a dense-bodied ceramic mold with all the usual hallmarks of a quality precision casting.

System Type/Construction: The machine has a build chamber with a movable base. A movable portal above the base contains a print head that contains the necessary amount of material and a roller that precompresses the material.

The advantage of this process is that it permits great freedom of design; owing to its ability to produce internal ceramic parts it also opens up possibilities of casting hollow components.

Material/Build Time/Accuracy: Any pourable material can be used including, for example, titanium or Inconell.

A disadvantage is that internal areas, that is, the surfaces of the molds, are hardly accessible after the mold is finished and therefore they cannot be appropriately finished.

Very thin walls down to 0.2 mm thickness can be realized. According to the the producer accuracy of up to ± 0.05 mm are achieved, thereby meeting the casting tolerances GT9 to GT12.

In contrast to castings from stereolithography and laser sintering models which, in principle, allow only one casting per model, the DSPC process is significantly faster. Its greatest economical advantage is seen on those solitary occasions when only a few mold prototypes are needed.

Post-Processing: The further procedure is similar to that of classical precision casting processes. The mold is baked (sintered). Afterwards surplus powder that is not firmly attached can finally be removed. In the last step the mold is filled with liquid metal. After this has cooled down the ceramic mold is destroyed and the model is removed. The surfaces may be polished if possible.

Process-Typical Follow-Up: The process is meant to be used directly for precision casting. Therefore, there are no doubling techniques.

3.3.6 Laser Generation

Laser generation is a process in which the material is applied as powder or in the form of wire and then melted locally by means of a laser. The layer is generated by solidification as a result of heat conduction into the component. In contrast to the sinter process there is no powder bed. This is a one-step process.

Strictly speaking the term laser generation is confined to the powder-based process derived from the laser coating (see Sections 3.3.6.1 and 7.2.3), while the wire-based process was developed from the laser buildup welding (see Section 3.3.7.2)

This process is described in detail in Chapter 7 and is still in the research and development phase due to continual optimization of various parameters.

Classifying these processes is difficult because they have been on the market only for a short time or heve been introduced just recently. No firm insight is available as to whether they are significantly further developed than those still in the development phase.

The LENS process is already commercialized and is described in the following section. The CMB process is a hybrid process and also commercialized. It is described in Section 3.3.7.2. The SLRP process is still in the development phase. It is described in Chapter 7 and the Arnold process, which is not really a rapid prototyping process, is described in Chapter 4.

As all these processes work with material suitable for tools there are overlaps with Chapter 4.

3.3.6.1 Laser Engineered Net Shaping (LENS) – Optomec

LENS 750 / 850

Optomec Design Company, Albuquerque, NM, USA

Short Description: The LENS process applies the powder by means of a laser to the locally molten area of the component. In this way a component is generated line by line and layer by layer [KRE98]. The process is identical to the well known laser generation and resembles basically the process developed at the Fraunhofer Institut für Lasertechnik (Fraunhofer Institute of Laser Technology) (FhG-ILT) (see also Section 7.2.3).

Range of Application: Technical prototypes, rapid tooling.

Development State: Commercialized since May 1998. According to the producer, in 1998 four systems were sold in the United States.

History: Optomec was established in 1963 by Thomas Swann. The company supplies primarily optomechanical systems to research institutions, universities, and as accessories to rapid prototyping installations. The LENS technology was invented and developed by David Keicher at Sandia National Labratory (SNL). He became a partner at Optomec in 1997.

Development Partners/Strategies: The LENS process was developed and patented by Sandia National Labratories (SNL), USA. LENS is a registered trademark of SNL. Optomec took over the LENS process in license from SNL in 1997 and further developed it into an industrially useable system. The strategic target is the direct production of metal functional prototypes and series-identical components using various metallic and ceramic materials. A development partnership exists with SNL.

Data Formats/Software: The LENS system reads error-free STL data and CAD data via the usual neutral interfaces. The system also functions with Solid Works Rapid Prototyping software (optional).

The system software includes all elements necessary for the preparation and control of laser generation processes.

The Principle of Layer Generation: The powder is fed into the focus area of the laser by several nozzles arranged around the laser beam. It is incipiently melted and so applied to the component. The material solidifies after cooling as a result of heat conduction into the component.

The model is generated line by line (raster process) while the line structure is turned by 90° from layer to layer. Nothing is known about further scanning strategies or about the methods by which the part is fixed on the positioning unit. A 700-W (cw) Nd:YAG solid-state laser with power control is used for the melting. The laser can be upgraded to 1.4 kW.

The producers claim that the laser process causes a refinement of the grain in the resulting structure, giving the LENS-generated parts a higher tensile strength and a higher ductility than those of the materials themselves [KRE98].

The build process takes place in a completely sealed process chamber filled with protective gas appropriate to the material used. It is equipped with a large window for observation purposes, a gas circulation device, and a video camera.

System Type/Construction: Optomec supplies a complete system consisting of software and hardware for the processing of CAD data and for the process control as well as the process chamber itself, sealed against environmental influences and working under protective gas, the laser, the coating unit, and the build platform (Fig. 3-76).

The x-y contour of the component is assured by a positioning system with brushless servo-drive. The dimensions of the process chamber are at least x, y, z = 250 · 250 · 250 mm. The direct control of multiple powder nozzles is possible. The laser coating head remains stationary in relation to its beam axis and is moved after each layer by one layer height in the z-direction. From the start the construction process, which involves moving the weight of generated component, is restricted to relatively small components if a reasonably high dynamic is to be achieved. Although the process takes place under various protective atmospheres it runs under normal atmospheric pressure and approximately at room temperature.

Material/Build Time/Accuracy: According to producer specifications, absolutely dense components are generated from various materials, among others stainless steel, titanium, and special alloys, which demonstrate the physical-technological properties of series parts.

The LENS part shown in Fig. 3-36 measures approx. 50 mm at its widest point and needs approx. 3 h for construction. Accuracy are not specified, but the photo shows irregularities and meltings within the layers as well as around the drillings so that an accuracy higher than ± 1 mm cannot be expected according to this estimate. The producer specifies a tolerance of 1/10 in relation to the dimensions of the finished part.

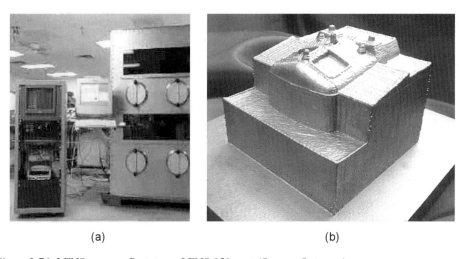

(a)   (b)

Figure 3-76 LENS-process: Prototyper LENS 850, part (Source: Optomec)

The LENS process has the fundamental advantage, theoretically at least, that components can be built of different materials. According to the producer, an accuracy of 125 microns (0,3 mm) in the x-y direction and of 250 microns (1 mm) in the z-direction is specified for such a small component. It seems possible to provide the components with supports of copper both for external parts as well as for internal hollow spaces. Further information is not available.

Post-Processing: If necessary removal of supports, polishing. Nothing is known about thermal processes.

Process-Typical Follow-Up: None.

3.3.7   Conventional Prototype Processes and Hybrid Processes

For the optimal use of rapid prototyping processes it is important to know exactly the possibilities and the limits of generative production technology and to take this into account when choosing one specific prototyping process. Like all processes, rapid prototyping also covers only one process window and is jostled by competitors from all sides. The list of these competitors is long and it is difficult to define exactly where they stand among themselves and vis-a-vis rapid prototyping. The expression "conventional processes" as opposed to "rapid prototyping" processes remains valid only so long as these are new. The processes posing the greatest competition for generative production processes in certain aspects and that are referred to here as conventional processes are listed in ascending order according to their degree of competition to rapid prototyping:

- Sawing

- Drilling

- Turning on a lathe

- Polishing

- Flex forming

- Hydro forming

- EDM

- Laser carving

- Winding processes

- High speed cutting (HSC).

The list shows that the transition between the so-called conventional and the so-called rapid prototyping processes is in a state of flux. Winding processes are exceedingly well suited for certain components of basically rotation-symmetrical form. Because they work without tools on the one hand and can build up the geometry additively on the other, they could be considered rapid prototyping processes. Compared with most rapid prototyping processes the winding processes have two great advantages: they work with series materials and they are technically and economically suited for production processes. Although other processes such as flexforming and hydroforming need tools they also can be considered as follow-ups to rapid prototyping processes as such tools are relatively easy to make, and they can also be made by rapid prototyping processes.

Conventional processes and their capabilities, potential, and limitations are not discussed in detail in this book. A number of excellent publications are available that discuss these matters from various points of view. In a publication by Marschall Burns, conventional processes and their operational potentials and limitations are discussed under the aspect of automated production [BURN93]. Reshaping, powder metallurgy, and casting processes for near net shape production are compared with rapid prototyping in [KUN97].

3.3.7.1 Conventional Processes: High-Speed Cutting (HSC)

From a number of competing processes, one development – high-speed cutting (HSC) – is selected for discussion in detail because, first, it has the largest competitive potentials as seen today and, second, it is the basis for so-called hybrid processes enabling a combination with generative processes.

These high-speed milling processes, which became known as HSC (high-speed cutting) or HSM (high-speed machining), appeared some years ago as if from nowhere. In actual fact, however, 40 years have passed since their first technical realization. Two developments were basically responsible for their fast ascendancy to the strongest competition for rapid prototyping :

- During the last few years high-speed machining, formerly used for the machining of light metals in series processes, has developed so far that practically the entire range of materials can be processed by HSC machines today. This development led to HSC machines being used for making tools and molds.

- Rapid prototyping processes progressed under the cloak of rapid tooling to become effective tools for tool and mold manufacture.

Competition exists today in the area of tool and mold making.

High-speed processing has significantly improved the key parameters for machining. Figure 3-77 shows the basic course of metal cutting capacity per minute, surface quality, cutting forces, and tool life versus the cutting speed.

The rate of metal removal and the surface quality increase with increasing cutting speed; the cutting forces decrease. The only negative point is the decreasing tool life. This is not necessarily always as negative as the graph may imply when the higher advance speed is taken into account. The most important advantages are that, with increasing cutting speed not only do the cutting forces decrease but also the process heat is completely dissipated with the chippings. This has a positive effect when processing components with thin walls and such components that tend to warp if temperature gradients become too steep. It is possible to generate excellent surface qualities that show R_A values of 0.2 μm and R_Z values of up to 3 μm. These properties of HSC are the most important for tool and mold making. The production times, which can be very significantly reduced, are much more important in industrial (mass) production than in tool and mold making.

Because the prices for high-speed cutting machines and prototypers are almost the same, competition between them increases. It should be taken into account, however, that prototypers can still be used as "stand-alone solutions" except for the limitations discussed earlier, whereas a high-speed cutter machine requires a fully equipped mechanical workshop.

Figure 3-78 shows the example of a forging die (x, y, z = 180 · 90 · 52 mm) of tool steel (hardness 51 HRC) made by HSC technology. The component was produced within 3½ h with a cutting speed of 5000 mm/min and a spindle speed of 35,000 min[-1]. The roughing and planing is done in one clamping. For clearing, tools with cutting plates were used and for preplaning and planing coated carbide tools were used.

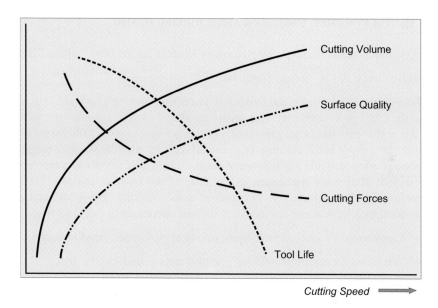

Cutting Speed ⟹

Figure 3-77 Graph of the key parameters of high-speed processing versus spindle speed [SCHU96]

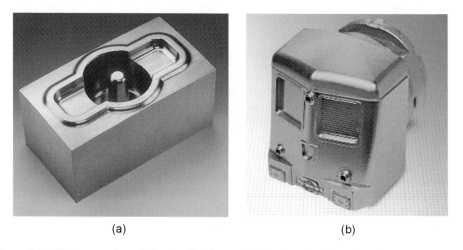

(a)   (b)

Figure 3-78 HSC processing: a) Forging die (Source: RödersTec); b) EDM electrode (Source: Kern)

For a detailed description of the technology of high-speed processing please refer to [SCHU96].

3.3.7.2 Hybrid Processes: Controlled Metal Buildup (CMB)

CMB – Controlled Metal Buildup, Supplementary Module for the HSC Milling Machine

A. RödersTec GmbH & Co., Soltau, Germany

Short Description: Controlled metal buildup is a selective laser buildup weld with wire. By means of the CMB process thin welding seams are placed closely together on a work piece in such a way that thin layers are generated. For this purpose common filler wire is melted on in layers by a diode laser. Because of the small melting pool the warping remains low. This process is integrated into the high-speed cutting machine of Röders as a supplementary module so that after each application the surface of the layer is milled. This process generates layers with parameters as is common in tool making. The CMB process was primarily developed for fast and economic repairs and alterations to tools and forging dies.

Range of Application: Functional prototypes, technical prototypes, rapid tooling.

The molds are suited for injection molding, pressure casting, and forging dies.

Development State: Commercialized.

Development Partners/Strategies: The CMB was developed in close cooperation between RödersTec and the Fraunhofer-Institut für Produktionstechnologie in Aachen (FhG-IPT) (Fraunhofer Institute of Production Technology, Aachen, Germany). The process is designed as a retrofitting supplement for Röders' high-speed cutting machine and is used preferably for the alteration of tools and their repair but is also considered as an alternative for the production of tool inserts, especially if a few fine details would require a high amount of cutting in a conventional process.

Data Formats/Software: The machine reads complete and error-free STL files. The manufacturer supplies software, which calculates the track geometry and the control data for the laser and wire feeding, analogous to the commercial CIM programs that were designed for milling.

The Principle of Layer Generation: Using a diode laser, the tool surface is incipiently melted point by point while a common welding wire is dipped into the molten mass by means of a wire feeding unit; material is applied in this way. A layer is generated by laying several welding seams next to one another. The low heat application by the laser together with the small melting bath ensure that the distortion is practically unmeasurable. After application, the contour of each layer is cut to the defined measurements by means of a high-speed cutter. Highly accurate, 100% dense, parts are thereby generated. Although the process was developed for alterations and repairs, complete new parts can be built with it as well (Fig. 3-79).

System Type/Construction: The CMB unit can be retrofitted in the common Röders high-speed cutter. For this purpose a movable slide is fitted in front of the cutting spindle on the z-axis. This carries the diode laser and the wire feeding unit. The slide is movable to enable it be pulled out of the work area during the cutting. The HSC milling machine is thereby not restricted in its performance and can be used at any time for HSC cutting as well as for the

CMB process. The regulations for laser safety require a complete encasing of the entire installation (laser safety class 1). The glass panels are therefore replaced by metal sheets except for one laser-safe observation window.

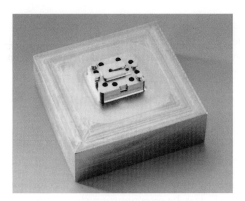

Figure 3-79   Injection mold generated by the CMB process (laser build-up and HSC cutting). Hardness 52 HRC. (Source: RödersTec, D)

Material/Build Time/Accuracy: The process resembles the classical laser buildup weld. There is a relatively wide choice of material because, in general, all common weldable wires can be used. The area welded on is 100% dense. According to producer information, reproducible pore-free surfaces can be achieved. The hardness is > 50 HRC. Because of the refinishing work with the miller, the accuracy of the applied area is equal to the typical machine accuracy of the HSC machine.

Post-Processing: None.

Process-Typical Follow-Up: None.

3.3.7.3 Comparison of Rapid Prototyping Processes and Conventional Processes

Most users of rapid prototyping processes, especially those new to the technology, are looking for some kind of system that allows an optimal comparison to be made between conventional and generative processes and that is tailor-made for their special needs. Such a matrix, which would assist in decision making, preferably something like a cookbook, does not exist and it is not advisable to try and put one together. Instead, the following discussion attempts to sharpen the perceptions toward the possibilities and limitations of generative processes in comparison to conventional processes, thereby fostering the ability to judge correctly in each specific case.

Materials

Conventional processes work with material or semifinished products that are optimally matched to the desired product in their mechanical-technical properties (product-oriented choice of material).

Rapid prototyping processes work with few basic materials whose properties are optimized for the process and therefore are less well matched to the product. Depending on the intended use, the material is roughly selected from a relatively small range of materials (process-oriented choice of material).

Tools

Conventional processes work with various tools that are optimally matched to the one particular task in hand. Sometimes they are even specially produced for the purpose and may have to be changed frequently (product-oriented choice of tools).

Rapid prototyping processes work without tools, that is, they do not use tools that match the particular model. The "tool" is a layer-contouring (forming) element that is not changed during one model-making process or from model to model.

Model Construction

Conventional models usually consist of several parts joined together that are often made of different materials, joint or assembled.

Rapid prototyping models are usually in one piece and of one kind of material. Indeed, they can consist of several model parts also but this is usually due to the limited size of the build chamber and not to the required model properties, or it is necessary for simulating the assembly of series.

CAD Compatibility

CAD models can be converted directly into processing programs for machine tools by means of corresponding programs (NC models). The volume of machine-specific information is so colossal that the processing of a program usually has to be done on the machine for which it was written (machine-dependent programming).

Rapid prototyping processes are developed from the CAD data record as standardized STL data models. All rapid prototyping processes known at present and machines can therefore be selected on the basis of the STL data record. This enables the optimal process to be selected without having to generate the data sets again (machine-independent programming).

Accuracy

Conventional processes usually achieve the machine-typical accuracy with the first processing run.

Rapid prototyping processes depend heavily on calibration. Optimal results are in general obtained only if the model is built several times or, at least, "added on" while the resulting divergencies are used for calibrating the machine.

Such calibrations can be reduced to a minimum if the machine is constantly used and experience is gained.

Influence of Manual Work

Precise conventional processes such as high-speed cutting or eroding result in surfaces that either need no post-processing at all or whose post-processing is mainly restricted to surface effects not causing measurable dimensional imperfections.

Rapid prototyping models create process-bound stair steppings that are comparatively easy to plane away manually because of to the relatively soft material. In this way, excellent surface qualities are achievable manually. However, dimensional imperfections are thereby created that can be a problem should one batch of rapid prototyping models be manually finished by several workers, for example when producing master models for precision castings. From this aspect, rapid prototyping processes are dependent on the qualifications and the day-to-day circumstances of employees which are reflected therefore by certain variations in their geometrical parameters.

From the preceding it can be clearly seen that conventional models certainly have advantages over rapid prototyping processes regarding accuracy, material properties, reproducibility, and, depending on the geometry, also speed. This is especially true when the geometries are relatively simple. Conversely, it follows that one of the most important demands for the successful use of rapid prototyping models is the following:

Rapid prototyping processes are best used for very complex components (which are needed quickly).

Even when keeping some of the positive aspects of conventional model making in mind, the danger of forgetting one of the basic advantages of rapid prototyping processes should not be overlooked:

Rapid prototyping processes ensure a consistent, uniform, and retrievable data record at all times for all project participants and are therefore an integrated section in product development strategies.

3.3.8   Summarizing Evaluation of Rapid Prototyping Processes

The desire of the user to have a simple-to-use and meaningful process at his disposal that assists him in choosing the most suitable rapid prototyping processes for any specific requirement in a fast and reliable way is as old as rapid prototyping itself and becomes even more acute the more new processes are introduced.

Although it was still possible 8 years ago, when the first edition of this book was published, to make direct comparisons between industrially useable machines on the market at that time (because they produced parts of approximately the same size with comparable degrees of detail within comparable price categories), this is in general no longer possible today nor would it make sense to do so. To compare the build chamber of a FDM Quantum with a microstereolithography installation, or the costs of material for stereolithography resins used in a SLA7000 with those of materials used for the Z-Corp machine would be meaningless.

The following paragraph will try to provide a sharpened perception toward the questions of which process is useable under what circumstances, which alternative processes would also fulfil the requirements, and which processes in special cases are less suitable.

For this purpose we first take a look at the criteria of model accuracy and surface quality. Thereafter, we take a look at benchmark parts and user parts.

We then try to classify industrial rapid prototyping processes by assigning them to model classes, thereby defining preferable fields of use. This is done primarily for functional prototypes in the product development phase "design."

3.3.8.1 Model Accuracy

When considering model accuracy we should differentiate between machine accuracy and process accuracy:

- Machine accuracy is the achievable accuracy under ideal circumstances, as specified in producer information (see also Appendices A2-4 to A2-14).

- Process accuracy is the accuracy resulting from taking the entire process chain into account.

For the user, the accuracy of the resulting model is the more important. This results as process accuracy from error propagation over the entire process chain. Machine accuracy is thereby only one element. Errors in the data model; the processing of imperfect data; and errors arising during data transfer, during model preparation, and during post-processing and finishing must also be considered.

The sum of all these influences cannot generally be expressed in numbers. Therefore, the influences will be discussed in an attempt to give the user a sense of for those aspects that are most important for achieving a high-quality model.

The achievable *machine accuracy* is limited by three main influences:

- The machine and its design features

- Principle of layer generation (process)

- Material

These aspects in relation to single prototypers have been discussed previously for each type of machine. In the following paragraph these prototypers are compared with one another.

Machine: Machine accuracy depends decisively on the dimensions of the sphere of action the shaping element has (laser beam diameter, droplet diameter, nozzle diameter, etc.) and on the method for controlling this element (scanner, plotter, etc.). The active diameter of the shaping element defines the basic accuracy while the scanner/plotter unit can introduce additional inaccuracy into the system.

Unlike plotters, scanner units cause additional angle errors so that varying accuracy may result depending on their position on the build plane. In addition, and this applies especially to fast scanner units, dynamic effects (overshooting) must be taken into account that are often also geometrically dependent (shadowing).

Principle of Layer Generation: When using plotter units that move the shaping element on an x-y slide across the build plane, it can be assumed that any inaccuracy they cause does not depend on the position. Only dynamic effects caused by curves, corners, and reversals should be taken into account.

Scanner units are faster not only at contouring, but they are also much faster at positioning than plotter units. This is important for the implementation of build strategies in which alternating model areas are exposed to avoid warpings. Because warpings, which result in model inaccuracy, also lead to unsatisfactory models, they are tantamount to inaccuracy caused by the machine.

Material: The quality of reproduction and the mechanical-technical properties of modeling material, together with the basics of model making and the consideration of specific prototypers, have been discussed in detail. For a comparable discussion, please refer to Section 2.5. The achievable model accuracy are inseparably connected to the material properties. Therefore, the material properties must also be taken into account when calculating the achievable accuracy (Appendix A2-15 to A2-25).

To assess the influences of single process elements on the ***process accuracy*** the most important elements of preprocessing, the generative construction process, and post-processing are the following:

Preprocessing

- Quality of CAD modeling

- Loss of quality caused by interfaces and data transfer

- Support structures and positioning the part in the build chamber

The aspects of preprocessing determining accuracy were discussed in detail in Section 3.1. The need to standardize the data transfer and the interfaces should only be reiterated here. Thereby it can be avoided to have to deal with problems in prototyping that basically are not prototyping problems at all.

Generative Construction Process

- Build parameters

- Material influences

Particularly in Sections 3.3.1 and 3.3.2, the generative construction processes were described in detail using the examples of stereolithography and laser sintering.

In addition a specific problem regarding prototyping is to be noted here: In most technical applications, the models have a dominant direction in which it is especially important to keep to the required accuracy while for other directions the required accuracy are of secondary importance.

Post-Processing

- Number of process steps

- Quality of post-processing and finishing

All elements have been described in detail in Chapter 2 and in the relevant machine-specific sections of this chapter and are therefore not discussed here again.

The Human Factor

It is especially important to be aware of the human factor in the rapid prototyping process chain. Descriptions of rapid prototyping processes often give the impression that these are model-making processes that run automatically without human supervision.

All currently used processes are highly dependent on experienced, skilled staff. This is valid for all process steps:

- Manual interventions when preparing the CAD data for the construction process and, if necessary, for the supports.

- Optimal adjustment of the machine. A model may have to be built until it has reached an assessable size enabling the machine to be calibrated accordingly.

- In two-step processes (post-curing), the influences of this second step must be taken into account. In a stereolithography process, for example, it should be considered whether the supports are best removed before or after the post-curing.

- The careful handling and the expert use of the material. Careful handling includes proper storage conditions for the material (moisture, light) and also correct recycling of reuseable material. In fact, most materials used for stereolithography and sintering go through a certain aging process if they have been in the machine several times even without having been directly used in the process. Certain quantities of new material should be added so that the desired material properties are maintained.

- Optimal positioning of the model in the build chamber with regard to required accuracy, taking into account the varying properties of the material when used in different building directions, and the economic usage of the build chamber (see Fig. 2-12).

- The significance of the human factor is obvious in the areas of model cleaning and finishing processes such as polishing, drilling, impregnating, varnishing, and so forth. Here, manual skill is required and an appropriately equipped workshop is also absolutely essential.

3.3.8.2 Surfaces

The largest disadvantages of rapid prototyping processes compared to conventional machining processes are the significantly rougher surfaces and the stepped surfaces in the z-direction.

Whereas these qualitative statements are considered to be exact, quantitative statements are relatively rare. Therefore, surface qualities were measured, for example, on the digger arm (see also Section 5.1.4). Those same parts were examined after they had been made by a stereolithography process (3D Systems, acrylic resin) and by selective laser sintering (3D Systems/DTM, polyamide). Each time, a measurement was made in the horizontal direction and another in the vertical direction at right angles to the first [VAN95].

The average peak-to-valley height R_A was established, which is defined as the arithmetical mean of all the values of the roughness profile within the measured length A roughness profile (R profile) for the measured stretch results. This is the high-frequency part of the surface profile resulting within the measured stretch after the elimination of the gradient of the compensation lines and after the waviness has been filtered out.

Owing to their layered construction the surfaces of generatively built components are structured regularly. For this reason the maximum wave depth W_T, representing a characteristic value of the waviness profile, is specified in addition to the specification of the average peak-to-valley height R_A. The waviness profile (w profile) is the low-frequency part of the surface profile which results within the measured length after the elimination of the gradient of the compensation lines and after the roughness has been filtered out. The maximum wave depth W_T is the vertical distance between the highest and the lowest point in the waviness profile. The graphs shown in Fig. 3-80. indicate the so-called P profile which represents the surface profile with roughness and waviness but after the elimination of the gradient of the compensation line.

In Fig. 3-80 and in Table 3-2 eight graphs are shown. Roughness values were measured and also the waviness of horizontal and vertical surfaces of stereolithography models with layer thicknesses of 0.125 and 0.250 mm and of selective laser sintering models with a layer thickness of 0.127 mm.

The influence of layer thickness in stereolithography is very small and can be interpreted only as a slightly excessive scan structure caused by an increased curing depth if this should not just be taken for scattering. This cause must also be attributed to the differing roughness and waviness of the horizontal plane compared with the vertical plane in stereolithography: The STAR-Weave technique puts layer upon layer of netted structures on top of one another whereby the mesh is not cured during the scanning (option: no fill). These can be found in the measurements as a result of their loosening during extended construction processes or during cleaning. The vertical planes do not show such structures and are therefore better by a factor of two regarding the roughness and by a factor of eight to nine regarding the waviness.

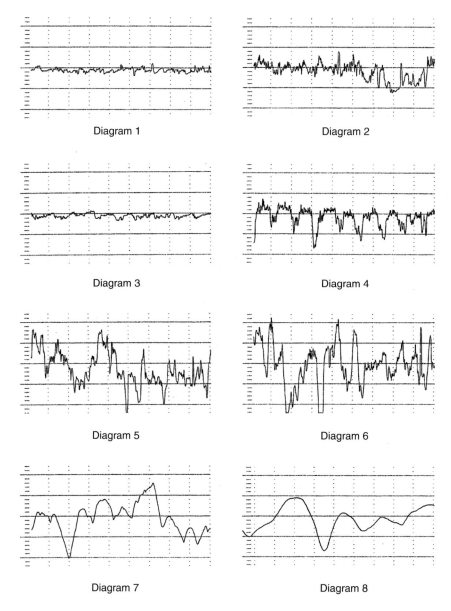

Figure 3-80 Roughness (R_A) and waviness (W_T) of the surfaces of SL and SLS parts (see table 3-2)

Table 3-2 Surface roughness (R_A) and waviness (W_T) of two equal components, one built by the SL technique (3D Systems, acrylic resin) and the other by the SLS technique (DTM, polyamide)

	Surface roughness (R_A) [mm]	Waviness (W_T) [mm]	Graph no. (Fig. 3-80)
Horizontal plane			
SL (0.125 mm)	6.44	93.20	2
SL (0.250 mm)	7.60	87.00	4
SLS (0.127 mm)	16.70	147.50	5
Vertical plane			
SL (0.125 mm)	3.13	10.40	1
SL (0.250 mm)	3.49	11.40	3
SLS (0.127 mm)	43.70	140.00	6
SLS glass blasted	24.50	111.00	7
SLS manually grinded	5.70	107.00	8

With selective laser sintering on the other hand, the horizontal areas are better by a factor of two than the vertical areas. In addition the waviness is overall significantly higher but is the same for horizontal and vertical areas. While this is attributable to the scan strategy (raster process), the greater degree of roughness on vertical areas is explained by the fact that these end in the powder bed and neighboring particles can be melted on by heat conduction; the significantly smoother horizontal areas end in the nitrogen atmosphere.

The last two measurement drafts show the influence finishing procedures have on the surface of sintered models. Draft 7 shows how the surface (roughness) can be improved by half again by glass blasting while waviness is not so much influenced thereby (significantly < 20 %). Intensive manual polishing achieves surfaces similar to those of stereolithography processes (roughness). Waviness is hardly influenced positively by polishing, probably because the process is manual.

3.3.8.3 Benchmark Tests and User Parts

Benchmark tests, or so-called user parts, provide sufficiently exact information about the achievable machine accuracy of a system. Here, it is certainly legitimate not to question the results regarding the multiple calibration of a machine or the multiple construction of one specific part. On the other hand, the user needs to be aware that the achievable accuracy often require an effort that cannot be economically fulfilled in the daily routine. These processes are, however, well suited for the characterization of machine performance.

Benchmark Parts. When the first prototypers were introduced, models were designed and built simultaneously, usually by more or less independent third parties, for the sole purpose of proving the abilities of a prototyper in the reproduction of details. Various so-called benchmark parts were developed in quick succession that were used primarily for comparing different systems. The common disadvantage of all benchmark parts is that, even when the details are selected reliably, they are still not as universally useable as they are presented to be. Trade-specific features, the experience of the designer, and the knowledge of any weaknesses and strengths of particular machines often find their way into the design of those parts. In the end most benchmark tests do not provide a solid base for comparisons because important circumstances are not taken into account. Among other factors are all the influences on process accuracy and, for example, the questions: Who did the build? How often had the part been built? How regularly was the machine calibrated? Were there transport influences? How soon after completion of the build process were measurements taken? Who took the measurements? – and many more.

Taking a closer look at so-called ***user parts*** shows that they are much more informative. It can be assumed that these are parts that were specially designed to demonstrate the advantages of *one* specific machine. It can also be assumed that the machine was precisely calibrated and operated by qualified personnel to achieve the accuracy published by the producer. Of course the storage and handling of the material will have been carried out with the utmost care. Taking for granted that this applies to all producers, user parts are quite a good basis for comparison. Apart from the accuracy achieved the design features of user parts especially tell us something about the strengths and weaknesses of a machine.

Measurings of Accuracy. Complicated measurement and interpretation processes followed in the wake of the first comparative examinations, where only a few main measurements were taken. Today these examine more than 100 points by means of automatic coordinate measuring devices pinpointing the accumulated faults and their frequency of their occurrence. Graphs show (see also Fig. 3-81) what percentage of the measured points are within a given tolerance range. Such graphs show both the error distribution and, above all, the largest single error. Figure 3-81 shows that when using vinyl esther in an EOS STEREOS MAX an epsilon-95 value of < 66 μm results [SER95]. In a company brochure 3D Systems also published an epsilon-90 value of 65 μm [JAC94], so it can be assumed that, under optimal conditions, the machines achieve approximately the same good results. These reports should be seen before the background that in 1989 comparable models still had epsilon-90 values of 400 μm.

Because the speed is an essential characteristic feature of prototypers, it is also important to ask about the achievable accuracy and how reliably these accuracy can be achieved. It has to be questioned, therefore, whether accuracy are achieved, so to speak, in the daily run of the mill with occasional calibrations in the course of routine maintenance, or whether it is necessary to prepare the machine with an extensive calibration process for each build process to achieve a defined accuracy. Such specifications cannot be given in general and should therefore be checked by the user personally.

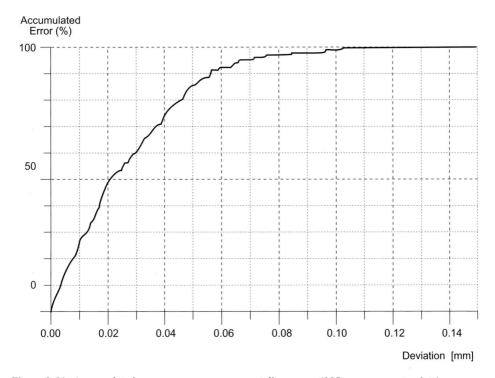

Figure 3-81   Accumulated errors versus measurement divergence (108 measurement points)

3.3.8.4 Comparative Evaluation of Prototypers for Functional Prototypes by Comparing Build Chambers, Accuracy, and Build Times

As discussed previously a comparison between all prototypers is neither sensible nor possible. Attempts at classifying prototypers according to the size of their build chambers, accuracy, and build times are therefore restricted to the majority of prototypers today that produce functional prototypes. The spectrum of parts includes telephone receivers, computer monitors, and toy cars as well as coffeemaking machines. Those parts that subsequently become series parts are produced preferably by the plastic injection molding process. Figure 3-82 shows a rough raster of the parameters: size of build chamber, achievable accuracy, and build time. Of these, the size of the build chamber is the only absolute value. Build time and accuracy depend on all kinds of conditions so that giving them numerical values was deliberately avoided.

From Fig. 3-82 it can be seen that LLM installations provide the largest build chambers followed by the FDM Quantum and the large stereolithography installations. Sintering installations and the small SL and FLM installations all have significantly smaller build chambers. FLM installations provide the largest spectrum of build chambers.

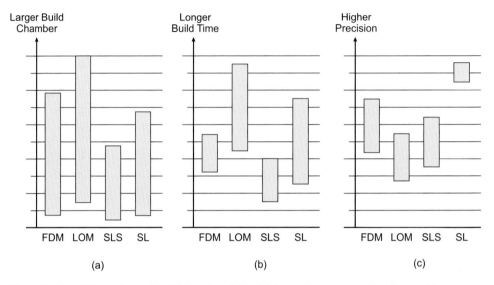

Figure 3-82 (a) Dimensions of build chamber, (b) Build time, (c) Accuracy of leading rapid prototyping processes

The peaks of the bars in Fig. 3-82a mark the largest machines available in the market; the valleys mark the smallest machines available. When comparing the build chambers it has to be taken into account that more models can be made simultaneously in one Cubital machine than in one large stereolithography machine of other producers. The reason for this is that, owing to nonexistent supports, the models can be nested into each other. Even these measurable specifications therefore cannot be taken at face value.

The achievable accuracy are manifested on two levels. From fig. 3-69c it can be seen that large stereolithography installations can work about twice as accurately as LLM installations which themselves are only 10% less accurate than sinter installations. FLM installations lie between these. In achievable build times, the sinter installations are clearly ahead of stereolithography, FLM, and LLM installations. This information was collected from various sources, for example, [AUB94]. Numerical values are deliberately not given. The descriptions are well suited, however, to giving a rough overview of the general properties of these four varying rapid prototyping installations.

It is clear that LLM installations should be used for the construction of large parts, for which the expectations concerning accuracy are not too high and relatively long build times are acceptable. Stereolithography installations achieve high accuracy in relatively short times in their large-volume build chambers. FDM installations achieve accuracy that lie between those of LLM and SL installations and can be considered as competition for SL installations.

However, this says nothing about the properties of the model. At this point the useable material should be taken into account. The LLM is restricted to paper and therefore produces woodlike models, although in principle, ceramic, plastic, and metal foils are available, or will be available, in the near future. Stereolithography uses photopolymers exclusively. The sintering process uses a wider range of materials that include metals and ceramics. The FLM processes include ABS but they are restricted to plastics.

The LLM process thus emerges as the first choice for large and yet quite detailed models of lower resilience. Sinter and FLM processes produce plastic models that are more detailed and resilient with finer structures and surfaces. Stereolithography processes are today preeminent in realizable accuracy and detail. Unfortunately, they are restricted to photopolymers. Functional models can therefore be produced only by casting processes.

Metal models can be produced only by sinter processes and by LLM processes that operate with metal foils.

3.3.8.5 Categorization of Prototypers in Model Classes

The interrelationships discussed are portrayed qualitatively in Fig. 3-83. In addition to single processes, the various model classes are shown. The widths of the bars in the graph indicate in which areas each model can be used successfully. Only entire plastic rapid prototyping processes are shown but not the results of follow-up processes.

The decreasing width of the bars with the growing series similarity of the model makes clear that all processes are less suitable for the production of series-identical prototypes. To improve the prototypes follow-up processes have to be used (see Section 3.4).

The bars of some processes, however, are growing narrower again with decreasing series similarity although no technical difficulties are to be expected. Out of economical considerations, in these cases it does not seem sensible to use a stereolithography model, for example, for the production of proportional models.

Overall, there are optimal ranges of use. Stereolithography models are especially suitable for design models; for functional models sinter processes are better used. LLM models are especially suitable for ergonomic and design models. 3DP models are best suited for proportional models.

This general categorization needs to be analyzed for each special case. Apart from the properties of single rapid prototyping processes discussed earlier the implementation of possible casting processes must also be taken into account.

There are many approaches to computer-assisted rapid prototyping selectors. Most of them are discontinued earlier or later because of the tremendous costs for maintaining and updating the system. The result of a rapid prototaping selector is usually not one proposal but several alternatives which are often subdivided into categories such as "well suited" and "partially suited".

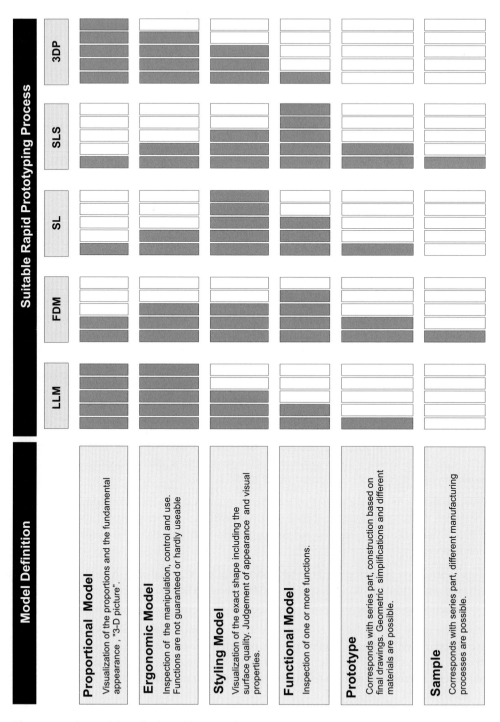

Figure 3-83 Categorizing plastic rapid prototyping processes and model types

It may seem disappointing that in most cases a rapid prototyping selector makes such a relatively rough choice instead of a clear decision. In fact, this reflects the difficulties discussed earlier, that of deciding between the pros and cons of different prototypers, and of deciding which one of them would be best suited for the task in hand. Whichever way you look at it, an entire store of process knowledge cannot be substituted by computer-aided selectors.

3.3.8.6 Development Targets

The weaknesses of current rapid prototyping processes as discussed previously can be apportioned to three groups:

- Rapid prototyping processes usually produce one model. In general, however, several models – usually 5 to 50 – are needed for production development, the simulation of assembly processes, and so forth even up to several thousand models are needed. Often, it is not economically feasible to build several models in prototyping processes as the unit costs hardly decrease as the number of units increases (depending on the process and the model size). If several models are produced the time advantage decreases or vanishes all together.

- Rapid prototyping models have relatively low-quality, stepped surfaces. The manual reworking is time consuming, especially if necessary for several models, and there is the additional danger of manually caused inaccuracy or faults which are different for different components.

- Rapid prototyping models consist of a process-typical material. Sometimes the mechanical-technological properties of the models differ significantly from those of the later series part. This applies even if the materials are nominally the same, for example, the mechanical-technological properties of standard 3D Systems and EOS polyamide materials largely resemble those of nylon (PA11). This is for the most part true also for single static and quasi-static values, although it should be questioned for dynamic strains for which no definite findings are yet known. Be that as it may, if the component is largely hollow and has countless tiny cracks it will be less resistant to permanent load bearing.

The weaknesses in the rapid prototyping processes of today point the way to the requirements of the future. Development targets are:

- Decreasing build time or machine utilization time

- Reduction or elimination of manual finishing procedures

- One-step processes and simplification or elimination of intermediate processes

- Improvement of surface qualities

- Reduction of variances in accuracy and surface quality between layer and build direction.

- Increasing of detail reproductions by means of thinner layers and by continual contouring in the z-direction

- Improvement of material properties and expansion of the range of material with the aim of production of series-identical prototypes

3.4 Follow-Up Processes

The weak points listed in Section 3.3.8.6 and the subsequent development targets are the subject of intensive research and development. Until industrially useable results are available the difficulties are bridged by using follow-up processes that are not based on rapid prototyping processes.

Rapid prototyping models should be considered as master models and, if series-identical properties are to be achieved, their surfaces need to be finished and they need to be molded in appropriate processes into geometrically identical models possessing the desired mechanical-technological properties.

The casting can be done in various processes depending on the target material. Plastic models are produced preferably with vacuum casting processes, metal models in a precision casting process.

The applied casting processes are not altogether new. With the growing significance of rapid prototyping new applications have simply been added. Experts for these casting processes often do not know how to handle rapid prototyping master molds. Conversely, users of prototyping are often unfamiliar with these casting processes. For a successful implementation therefore an intensive exchange of experience and if possible systematic experiments are necessary to master the new materials and the special requirements in handling them in a joint effort.

Molding processes produce positives (duplicates) as well as negatives (molds and tools). Molding process can therefore be considered a transition from rapid prototyping to the (rapid prototyping supported) rapid tooling. Rapid tooling is discussed in this chapter under the topic of molding processes. In contrast to the direct generation of molds and tools by rapid prototyping processes, the production of tools by molding processes is known as indirect (rapid) tooling. It is therefore discussed only insofar as is necessary for the understanding of the possibilities and limits of rapid prototyping in connection with molding processes.

Follow-up and casting processes are divided into those that are directed toward plastic components and others that are used for the production of metal models.

3.4.1   Target Material Plastic

Plastic processes are grouped as follows:

- Silicone rubber casting under atmospheric pressure (gravity casting)
- Silicone rubber vacuum casting [Room Temperature Vulcanisation (RTV)]
- Injection molding
- Photocasting
- Spin casting

While vacuum casting is of the greatest importance technicallly for the production of proto-types and miniseries, injection molding is most important for the production of series.

The simplest of these processes is ***silicone rubber gravity casting***. The surface of a rapid prototyping positive of any arbitrary material is finished. The positive is embedded in silicone material (silicone polymer) that, after solidification, is cut into two halves along a predefined parting line in such a way that the model can be taken out. Then, feeding tubes and vents are installed, the mold is closed, and liquid, self-solidifying plastic is poured in. After some time the model can be taken out. In one mold more than 10 second casts can be produced. The number depends basically on the complexity of the model, that is, whether the mold is damaged or not when the model is removed. The process is very simple, but the models have unsatisfactory mechanical-technological properties. They are in general not very resilient; they may be used as demonstration models and are therefore not very common.

Vacuum casting (RTV) is the most important of the prototype production processes. The process is identical to that of the silicone casting process as far as the mold making is concerned. The model is cast in silicone angles and this mold is then cut into two halves so that the model can be taken out easily. Then, the mold is prepared with feeding tubes and vents for use in a vacuum casting machine. There the mold is evacuated under predefined conditions (pressure, temperature, time), depending on the material, and, afterwards, filled with the appropriate vacuum casting resin. The advantage of vacuum casting is that the molds are evenly filled and the resin is degassed. Significantly better model properties result. Today a vast number of commercial plastic materials can be used as vacuum casting material, and the resulting models are comparable with the later injection cast mass produc-tion parts, at least with regard to certain selected properties (see Appendix A2-15 "RP-material and casting resins"). Depending on the vacuum casting material and the complexity of the mold, 10 to 30 – sometimes even 50 – models can be manufactured from one mold. If the mold is very complex, the rapid prototyping mold is usually damaged when the model is taken out. If it stays undamaged, more molds can be produced for the production of larger series. Figure 3-84 shows an electronically controlled vacuum casting machine basically designed as a vacuum chamber (Fig. 3-84, bottom). Figure 3-84, top, shows the two resin containers in the upper part of the vacuum chamber. In the lower part of the vacuum chamber the silicone mold is placed during the casting process.

Injection molding of components with thermoplastics is one of the most widely used production process in many industrial branches. Usually, molds of steel are used, sometimes of aluminum or bronze, which have to withstand the enormous mold clamping forces (up to several hundred tons), the corresponding pressures (up to 2500 bar), and the resulting temperatures of up to 400 °C or more over as short a cycle time as possible. A machine melts and injects the plastic into the clamped tool in one process. After it is sufficiently solidified the tool is opened and the injection molded part is ejected.

Preseries and small series have been produced successfully several times with metal rapid prototyping tools in injection molding machines and the process is considered to be techni-cally under control. Here, rapid prototyping tool inserts are fitted into steel master tools. The

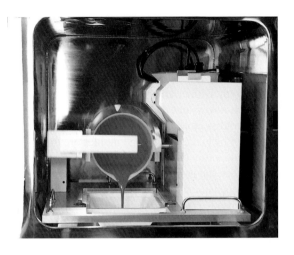

Figure 3-84   Vacuum casting installation: Process chamber (top), machine (bottom). (Source: MCP-
HEK)

tools are produced by molding as well as by rapid prototyping processes. This is a fast and economical way to injection molds for a pre series production of several hundred or up to a thousand prototypes. The use of rapid prototyping processes for the production of tool inserts is discussed in Chapter 4.

Photo casting differs from the silicone molding process only in the material used. The silicone mold is filled with a photosensitive resin and exposed through the transparent silicone mold. As in the stereolithography process, the part solidifies by photopolymerization.

Spin casting is an offshoot of the centrifugal casting process. The positive is vulcanized into rubber material and the tool is separated along a predefined parting plane. Several similar or different molds can be arranged concentrically around a rotation axis in one tool. The casting material is fed in on the inside so that the material is pressed into the mold by centrifugal force. Duroplastics and also the low-melting casting zinc alloy Zamak can be used, thereby connecting the spin casting processes with metal casting processes.

3.4.2   Target Material Metal

Although plastics are ubiquitous and are continually being applied in new areas, most functional parts in machine manufacture are still made of metal. Casting processes are of great importance in producing large series economically. One of its main advantages is the vast range of available alloys [SHE94].

This broad range of materials, however, can be used directly only in a few cases. In most cases the production must be done via molding processes. For this reason processes are used by which rapid prototyping models can be cast in metal, thereby facilitating an early test of functions and the planning of production tools.

The differences between the casting processes known today are manifested in the properties of the cast and the number of economically producible pieces. Up to now rapid prototyping supported applications have been available for precision casting, for special applications of sand casting, and for first applications of die casting. As regards rapid prototyping, all other casting processes are still in the development phase. Table 3-3 gives an overview and also shows the important development targets that must be achieved before rapid prototyping can be implemented satisfactorily in casting processes.

In characterizing molding processes, foundry science differentiates between the following:

- Permanent mold
- Lost mold
- Lost mold and permanent model
- Lost mold and lost model

Table 3-3 Classical casting processes and their demands on rapid prototyping

	Precision casting	Sand casting	Die-casting	Gravity die-casting
State of process	application	application	development	future development
Castings per year	1...100,000	1...1,000,000	> 50,000	< 100,000
Precision, details	high	medium/low	high	medium
Applicable RP process	SL, SLS	SLS direct application. LMPM-SLS	(SLS) metal sintering	LOM
Special casting processes	expanded polystyrene core casting	sand shell casting	-	-
Special RP processes	permant molds for making the wax masters	direct sintering of polymer coated sand	-	-
Comments	when melting the RP master: no of parts = one when using directly sintered ceramic molds: no of parts < 10	when melting the RP master: No of Parts = one	-	-
RP development goals	surface quality			
	mimizing remaining ashes	mechanical properties	reducing edge wear	-
	permant molds for making the wax masters: no of parts > 100	quick and cheap making of the mold	resistance against: pressure, temperature and temperature changes	resistance against: pressure, temperature and temperature changes

The expression "permanent" in connection with rapid prototyping should not be taken too literally, as rapid prototyping models and molds do not yet share the same levels of resistance to temperature, pressure, and wear as is expected of conventional (steel) models and molds.

The most important casting processes are matched to the characterization. They are listed in the order of their importance for rapid prototyping: precision casting, lost foam casting, sand casting, centrifugal casting, die casting, chill casting. Their interrelationships are shown in Fig. 3-85.

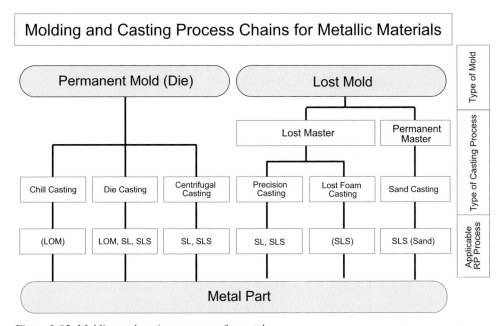

Figure 3-85 Molding and casting processes for metals

Permanent molds

Chill casting and die casting belong to the casting processes using permanent molds. In both cases divided metallic molds are used. In die casting, the casting metal is filled into the mold under high pressure and at high speed. Extremely small tolerances and high surface qualities can be achieved. In chill casting the casting metal is filled into the mold either by gravitational force or by a slight excess pressure against gravitational force ("low-pressure chill casting"). Internal contours and undercuts can be realized by placing a disposable sand core inside or by the use of a mechanically or hydraulically operated metal slide. The surface quality depends on the surface of the chill mold and is therefore superior to all sand casting processes.

Die casting and especially chill casting put extremely high demands on the mold and on its thermal and mechanical stability. Molds produced with the aid of rapid prototyping therefore have up to now, been used only in single cases and only for die casting. Mainly the materials are a limiting factor. Centrifugal casting still has no technical significance in connection with rapid prototyping

Lost Molds

Lost Mold and Permanent Model

Sand casting has the greatest industrial significance in this group. It is implemented in machine molding (industrial scale manufacture) and hand molding (small series). In these molding processes a model contour is reproduced by densification of sand around a model. The models are of wood, plastic, or metal, depending on the required life span.

Owing to their woodlike character, LLM models are a suitable replacement for classical wooden molds. The casting process then remains unchanged. The weigh only a few kilograms because of the size of the machine.

In shell mold casting the silica sand, which is coated by a thermoplastic, is applied to a heatable model contour. Because of the heat, the sand near the model solidifies, while superfluous sand can be used again. The shell molds generated in this way are joined together and stabilized by backfilling (e.g., by steel grit). There are no rapid prototyping supported approaches as yet.

A rapid prototyping alternative to the classical sand casting with polymeric-bonded sands is the direct sintering of molding sands. These processes are, however, preferably suitable for complicated cores. DTM sinters the sands in the plastics machine (see Section 3.3.2.2), for which EOS has developed its special product line (EOSINT-S) (see Section 3.3.2.3). This includes one machine that, in regard to build chamber and productivity (two laser scanner units), is matched to the size of internal combustion engines. The advantage of sintering molding sands lies in the ability to produce directly complex cores for cylinder heads, for example, which led to tremendous time savings in engine development. The disadvantage is that one sinter model is needed for each core (moving away from shell mold casting) and that the surfaces are relatively rough. This is usually acceptable for the cores; for the mold, the achievable surface qualities are often not acceptable, especially as with some geometry post-processing is difficult.

Lost Mold and Lost Model

Lost-wax casting and lost foam casting are two processes in which mold and model are used only once. Both processes allow a broad spectrum of geometries, as only one-piece molds are used and removal from the mold must not be taken into consideration when designing the mold. The precision casting process is the most important of the lost-wax casting processes. Here a wax model cluster is made first, then repeatedly dipped into ceramic slurry, and subsequently sanded with fireproof material, thereby coating it with a stable ceramic layer. After the wax is removed by melting, for example, and the mold has been baked, the cast is produced by filling the mold with liquid metal. A high-quality surface and the possibility to reproduce exact replicas of filigree contours are special features of the precision casting. The process is shown schematically in Fig. 3-86.

Production of Wax Samples (by Injection Molding)

Precision Casting Process

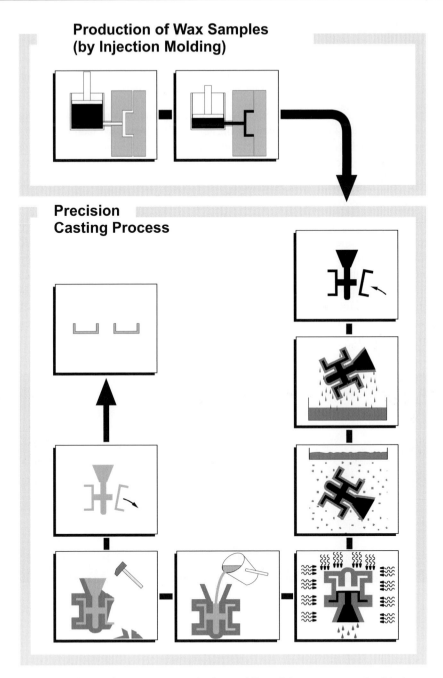

Figure 3-86 Precision casting process: (a) Injection molding of the wax masters (positive);
(b) Precision casting process (schematic)

Selective laser sintering and FLM facilitate the direct production of wax models that can be cast directly in a classical precision casting process without major process changes. Not so favorable, however, is the fact that these wax models break easily and also that the model is lost so that for every casting a new rapid prototyping model is necessary. Therefore, this process, if used for more than one piece, is expensive and slow.

For larger numbers of pieces, it is better to produce a permanent mold by molding rapid prototyping (preferably stereolithography) master models (see also Section 5.2.4) and to make the wax models for the precision casting process by injecting wax into those molds. The molds are then saved.

For direct precision casting processes with lost models (and as a consequence also lost molds) interesting alternatives arise by applying rapid prototyping processes.

As solid stereolithography models generate large amounts of ash when the precision casting molds are baked and their expansion can easily cause shell cracking, the producers of stereolithography machines have developed special build techniques to enable solid model parts to be built hollow thereby avoiding material accumulation. The Quick Cast process (3D Systems) has the longest experience in this sector and can show many impressive examples. Alternatives are "skin and core" (EOS) and "fast sculp" (F&S). Figure 3-87 shows a complex precision casting part (Fig. 3-87b) produced by the Quick Cast process (3D Systems, Fig. 3-87a).

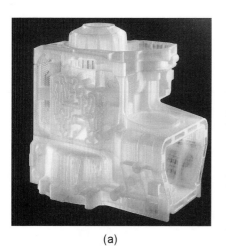 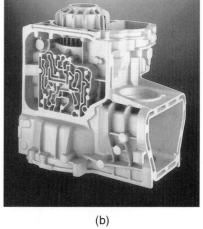

(a)   (b)

Figure 3-87 (a) Quick Cast model, (b) precision casting part. (Source: VW / Tital / Stemo Tec)

All molding processes demand master models with high-quality surfaces, not only because of the high demands made on the components but also to ensure an easy removal of the mold and to lengthen the life span of the tool. An intensive, usually manual, finishing of the surfaces and, depending on the process, the sealing of the areas near the surface is without exception essential.

Castable molds can also be produced by selective laser sintering. Polycarbonte (DTM), of which most examples have been published, and polystyrene (3D Systems, EOS) are especially suitable. Extremely high-quality and perfectly sealed surfaces are essential for high-quality castings. To this end, highly absorbent models undergo manual finishings and various wax-based infiltration processes. Using the example of a cast from a sinter component made from polycarbonate, Fig. 3-88 shows the effect a wax-sealed surface has on the result of the casting under otherwise identical conditions [DIC95].

Figure 3-89 shows a precision casted part part and the corresponding master model made by polystyrene laser sintering.

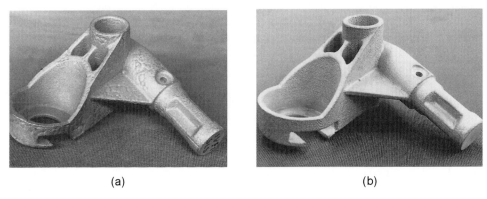

(a)   (b)

Figure 3-88 Precision casting of a polycarbonate model with (a) unsealed and (b) sealed surface.
(Source: EARP)

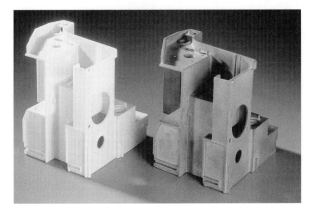

Figure 3-89 Polystyrene master model (left) and precision casted part (right). (Source: EOS)

The direct production of a (lost) mold is made possible by the Soligen DSPC process which directly sinters artificial resin-bonded ceramic powder with the 3D printing process. With this process whole mold clusters can be produced. Nevertheless, one rapid prototyping mold is necessary for every casting process, approximately one to ten pieces. The disadvantage lies in the fact that the internal areas of complex molds are not accessible for finishing procedures.

Within the lost foam process, the model stays in the mold after the molding and is burned by the liquid metal during the cast. Therefore, it is not possible to realize cores and mold drafts. There are two different production procedures. For the production of single models, the model parts are cut out (cutting, milling) of a slab of foam material (polystyrene) and then joined together with glue. The backfilling of the models is carried out mostly with self-curing molding materials (artificial resin-bonded silica sand). The production of models for series parts, on the other hand, is carried out with special foam moldings into which expandable polystyrene (PS-E) or polymethylmethacrylate (PMMA-E) is blown. The single models are glued together to form clusters. After the surfaces are finished (covering the surface of the model with a fire resistant coating) the models are backfilled with binderless sand.

Solids produced by rapid prototyping processes are at present the subject of research.

High-quality results cannot be achieved by high-quality rapid prototyping models alone. Good quality castings require direct cooperation between the foundry and the model maker to optimize all aspects of new model materials.

Numerical simulations have been used successfully for optimizing the casting process. With the aid of these simulation programs knowledge is gained at an early stage about thermal tensions, porosity, and the mold filling process and can be taken into consideration in the design of the feeding and gating system and in the choice of casting parameters. Rapid prototyping processes do not replace simulation processes. It is essential to optimize the total result by combining numerical and physical (rapid prototyping) simulation.

Simulation processes must be developed further for the special requirements of rapid prototyping materials and molds; this has not yet been achieved.

4 Rapid Tooling

This chapter is understandable on its own, provided a basic knowledge exists, thereby enabling a quick approach to the subject. Repetitions are unavoidable because of the nature of this concept, but these are kept to a minimum.

While the first applications of rapid prototyping processes focussed on the generation of 3D physical models according to functional prototyping practice, the idea of using the same technology for also making tools mainly for plastic injection molding developed early on.

The interested amateur is often confronted with euphoric success stories about "rapid tooling" in the literature and at conferences. This often causes exaggerated expectations ("Owing to this information the customers long for a tool made from nothing", according to Dana Zavettori, application engineer at AlphaCam, a Stratasys distributer in Gemany). But total rejection is also not unusual. Nobody should let his opinion be undermined by these extremes. The right decision is to examine the processes critically, assessing the advantages and disadvantages of often fundamentally different processes in view of the applications required, and be alert for new developments.

As mentioned in Section 1.4.2, rapid tooling is an application of rapid prototyping technology in the sense of a strategy. It follows that rapid tooling processes are identical to rapid prototyping processes or are based on the same principles. This is then known as direct (rapid) tooling. Indirect (rapid) tooling is defined as the production of tools by castings from rapid prototyping master models (Section 3.4). Generally both the direct and the indirect methods are often referred to as rapid tooling or tooling although they are two completely different methods.

This chapter is rather more extensive than the mere description of rapid prototyping applications for the production of tools and molds would normally require. Specific basic knowledge concerning rapid tooling is also discussed, as the intention of the user of tooling processes is different from that of the user who intends to use rapid prototyping for model making.

4.1 Principal Ways to Metal Tools

The term "tools" mainly refers to components that are used as permanent molds for plastic injection molding or as dies for metal die casting. When speaking of *rapid tooling* we inevitably mean the production of **metal tools** and **tool inserts,** or of tools that basically have the same properties as metals. Correctly within the Rapid Prototyping process there are made patterns which are used to make tools in a tool making shop.

The concentration on tools for plastic injection molding or metal die casting in this section does not mean, however, that the rapid tooling is specifically restricted to these processes.

Figure 4-1 shows the basic construction of a typical tool [GRU93] for a plastic injection molding.

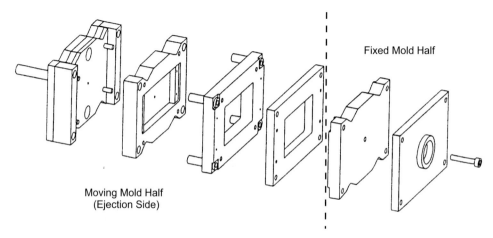

Fixed Mold Half

Moving Mold Half
(Ejection Side)

Figure 4-1   Tool for injection molding machines, exploded view. Picture taken from [GRU93]

All injection molding processes have in common that temperatures and pressures usually applied in the mold in mass productions are not reached with prototype tools and therefore neither are their flow patterns in the molten mass. The tools must be optimized, therefore, to match lower pressures and temperatures, wider flow diameters, a higher number of sprues, and the different flow pattern of the plastic. From the standpoint of classical tool making this looks rather like a step backward.

Rapid tooling is in fact merely an inversion (positive/negative) of data for otherwise unchanged design processes. In practice, however, there exist some complicating circumstances, the majority of which apply not only to rapid prototyping:

- The geometric data of a mold differ greatly from the inverted geometric data of the final product. Shrinkage and removal from the mold, ejector, inflow, venting, and cooling ducts, and so forth have to be taken into account. To define the geometry of a tool the appropriate software must be available and – above all – the necessary know-how for tool making. The software is based on the CAD model of the final product as is usually produced in the CAD draft.

- The majority of prototypers today are basically constructed for processing plastics. In most cases the available materials are unable to withstand the thermal and mechanical strains in an injection mold. The properties of metal materials are not identical to those of steel tools.

- Two-piece tools, consisting only of a mold top and a mold bottom, are rather unusual. Normally they are multipiece tools that are assembled with the aid of mold inserts, slide bars, and standard tools and that can be worked with the classical methods of tool manufacture and the single parts of which have similar mechanical-technological properties.

There are at present various processes leading to practical success in producing tools as fast as possible. They are roughly divided into two groups: those that derive from plastic rapid prototyping models and processes, and those that are based on the direct processing of metals (usually with lasers) and derive from welding or coating processes.

Because extensive experience has been accumulated with rapid prototyping processes based on plastic materials, it seems sensible to produce metal tools on the basis of plastic rapid prototyping processes. This indirect production is achieved by various molding processes (see also Section 3.4).

The direct production of tools with rapid prototyping processes (i.e., in the prototyper) requires first the adaptation of plastic rapid prototyping processes for the processing of metal materials. Two processes are commercialized.

In the first case, powders of metallic or other materials, for example, ceramics, are coated with plastic and are then sintered in the prototyper basically like plastic (indirect process, DTM/3D-Systems, Section 3.3.2.2).

In the second case, a mixture of high and low melting point metal powders is sintered directly in the prototyper which is basically the same as a plastic machine (direct process with multicomponent powders, EOS, Section 3.3.2.3). In a modified machine single-component steel powder possessing similar properties as cast steel can also be used.

In each case, useable metal parts are produced by (multistep) resintering and infiltration processes and elaborate finishes.

Metallic rapid prototyping processes in which tool steel is directly processed in the prototyper are on the threshold of commercialization (see Section 3.3.6.1 and Chapter 7). Development engineers pursue the rapid prototyping approach, that is, the direct sintering of single-component metal powders in a powder bed, as well as further development of laser processes for buildup welding (see Section 3.3.7.2 and Chapter 7) or coating.

All presently available rapid prototyping processes – and those that will be available in the foreseeable future – for the rapid production of tools and tool inserts are based on processes that include a phase transformation (liquid, solid), and they are therefore prone to shrinkage. The resulting variations in absolute dimensions are one of the main reasons why rapid tooling processes at present are limited to relatively small tools and tool inserts. There is also another additional disadvantage in rapid prototyping processes: the stair stepping of surfaces. Usually this results in a relatively rough surface, even when the effect may be minimized by thinner layers or by a variable layer thickness adjusted to the geometry.

4.2 Metal Tools Based on Plastic Rapid Prototyping Models

4.2.1 Precision Casting of Rapid Prototyping Master Models

All rapid prototyping models (positives) may be molded by precision casting (Section 3.3.2). In some processes, for example, in special stereolithography build styles, in laser sintering with polycarbonate or polystyrene and in extrusion processes with waxlike materials, rapid prototyping models may be melted out and are therefore directly useable in precision casting processes. With the aid of silicon molds, all (not just rapid prototyping) positives may be molded in wax and may then be used in precision casting. In this way tools and tool inserts, preferably of aluminum, may also be produced by casting.

In practice, precision casting has proved useful for the production of positives. For producing tools and tool inserts this process is rather seldom used.

4.2.2 Castings from Stereolithography Models

4.2.2.1 Direct Use of Stereolithography

Stereolithography is the most exact rapid prototyping process with regard to tolerances and surface quality. The direct use of stereolithography tools was long considered unfeasible in view of the glass transition temperature of the stereolithography material of approx. 75 °C. Nevertheless, systematic developments led to the ACES injection molding process (AIM). Stereolithography tools are produced aided by specialized stereolithography build style, ACES, and directly used in injection molding machines for the production of miniseries of up to approx. 200 pieces. Compromises are made, especially with regard to the cycle time which usually takes several minutes. This process is not of great importance, in spite of some successful applications [JAC97]. This could change with the trend to smaller numbers of pieces, as the process aims to bridge the economic gap between vacuum casting and injection molding. Therefore, it is classified as "bridge tooling." All the same, a comparison, published by the producer, between AIM tooling – 4 hours – and conventional tooling – 90 hours – does come out impressively in favor of the AIM process [JAC97].

4.2.2.2 Indirect Use of Stereolithography

The mechanical-technological properties of stereolithography materials in most cases do not fulfil the requirements for direct use in an injection molding process. Therefore, processes were developed of making tools by molding them from stereolithography models.

Counter Casting and Semigenerative Processes

Processes such as backup casting or the molding of stereolithography models according to the method of counter casting exist and are often used. For this purpose master models must be built with the required scalings. By counter casting with usually aluminum-filled artificial resins, comparatively simple and mainly uncooled two-piece tools are generated. They usually do not possess sliders. Undercuts are realized by means of loose parts (manually operated sliders). Depending on the component, several hundred or up to a thousand components may be cast in such molds (Fig. 4.2).

In a modified procedure based on the same process, it is assumed that free-form areas, typical for rapid prototyping, may be produced by rapid prototyping processes, preferably by stereolithography, and that surfaces should be molded by low-temperature metal spraying. All other elements of the tool, the ejector, and so forth are produced by machining processes or standard tools are used. A tool constructed in this way and inserted into a steel master mold is shown in Fig. 4-3 [GEB97].

3D-Keltool

The Keltool process was developed as the 3M Tartan Tooling process in the United States during the 1970s. In 1996 it was taken up by 3D Systems and further developed in view of the new possibilities for rapid prototyping. 3D-Keltool – the correct name since then – is based on the molding of stereolithography master models by long-term, low-temperature sintering of a special metal-plastic powder mixture (Fig. 4-4).

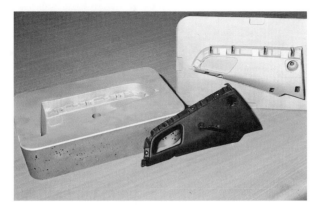

Figure 4-2   Tool cast made of aluminum-filled resin by counter-casting from a SL master model, injection molding part (mold material: EP 250 – MCP-HEK) part material: PA6 GF). (Source: Elprotec)

Figure 4-3 Semi-generative tool. SL model, coated with aluminum in a metal injection process, with milled inserts, steel master tool. (Source: CP)

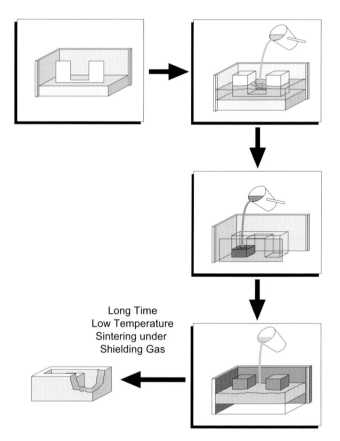

Long Time
Low Temperature
Sintering under
Shielding Gas

Figure 4-4 Keltool-process, principle

This process combines the advantages of stereolithography – high build speed and good surface quality – with the good reproduction properties of soft silicon molds. A high-temperature silicon mold is produced from a positive master model via an intermediate mold. This mold is then run in with the Keltool material consisting of wolfram-carbite and tool steel which then sets into a hard form. A single or double molding process is necessary depending on whether the positive or the negative is used as master model.

The process is insofar identical to the low melting point metal casting process (LMPM) in which the resulting mold is made of a low-melting metal alloy.

After solidification the powder mixture has its green part stability. In a reducing process with hydrogen (approx. 120 °C) the binder is expelled and the porous form is sintered. The hollow spaces are then infiltrated with copper. Details of the process itself are kept secret.

Keltool is used only by those few users possessing a direct license from the producer, who often produces tools to order. The first license holder abroad in Germany was Moeller GmbH. Being the producers of relatively small plastic injection components and electro-magnetic units assembled therefrom, Moeller uses the Keltool technology mainly for the production of tools that are directly molded with the target material. Figure 4-5 shows as an example a silicon mold and its corresponding stereolithography master tool; Fig. 4.6 shows the tool [DAA98].

Figure 4-7 shows the two mold parts for another component, a shot still in the mold and the finished product.

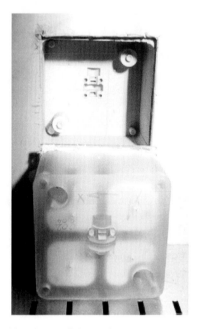

Figure 4-5   Keltool: silicon mold and stereolithography master. (Source: Moeller)

Figure 4-6 Keltool: tool insert with silicon mold. (Source: Moeller)

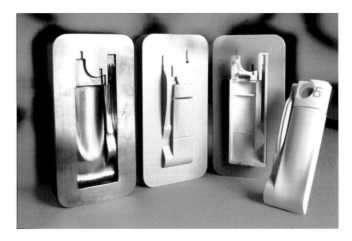

Figure 4-7 Keltool: tool with shot and product. (Source: 3D Systems)

The tolerances of ± 0.2 % result in acceptable accuracies for small parts. The hardness is specified as 44HRC, which is acceptable but lies significantly below the 50HRC mark. The tool inserts can be worked with the classical methods of tool manufacture. The possible use of the tools includes preseries, series kickoff and, depending on the batch size, even series.

Production time from the generation of the stereolithography master model through to the application of the 3D-Keltool insert is approx. 10 days and is significantly longer than for the various metal sintering processes, but it produces better quality surfaces and finer details.

Depending on the complexity of the geometry this process is significantly time-saving – in single cases by up to 30% – especially in comparison with EDM processes. The time advantage of 25% to 40% claimed by the producer should not be misunderstood as a process property but must rather be considered as a geometry-dependent standard value.

A variant of the process enables the use of copper-wolfram-carbide alloys for the production of electric discharge machining (EDM) electrodes with surfaces of up to $R_A = 0.4$ µm.

In accordance with the principles discussed in the preceding that the production of tools is basically an inversion of the (positive) component the Keltool process may be used for both for producing a positive (component) as well as a negative (cavity).

Other Processes

In the course of improving quality and the promulgation of rapid prototyping processes, "new" rapid tooling processes also enter the limelight. This is valid for generative processes and process chains with molding processes alike. According to Terry Wohlers (8/98), an expert on international rapid prototyping, "the list of rapid prototyping processes is continually growing. Examples are CEMCOM, Dynamic-Tooling, express-tool and Extrude-Hone. Each approach is connected with special restrictions, but each promises to reduce production times for molds and mold inserts. None of these processes may be considered completely commercialized at present and the development engineers are strangely cautious when asked how their processes work." [WOH98]. This statement signifies the tendency apparent in the entire rapid prototyping world of bringing new developments onto the market too early, thereby turning customers into alpha users. In view of the extremely rapid development such statements must always be questioned. To date, not only the indirect approaches as mentioned in the preceding, but also the direct rapid tooling system by Extrudehone have been developed further (see Section 3.3.5.2).

4.3 Metal Tools Based on Plastic Rapid Prototyping Processes

The first commercialized rapid prototyping process for metal tools was DTM's (now 3D-Sytems) Indirect Selective laser sintering process (about 1993) which is known today as "RapidTool." The process is available in two combinations of materials: copper-polyamide for low-resistant plastic injection casting processes for miniseries, and stainless steel-bronze (RapidSteel) for higher resistance and a larger number of pieces and also for metal die casting. The metal powders are coated with plastic so that in the sintering machine only the plastic coatings are fused together.In this respect the machine resembles a plastic sintering machine. A rather fragile so-called green part is generated from which the binder is removed in a multistep process; this is then resintered and infiltrated with copper or polymers.

The process takes more than 40 hours and includes heating and cooling so that special care has to be taken to avoid warping. With this process comparatively thin walls, undercuts, or other complex geometries can be reproduced that are difficult to produce in milling processes. From several hundred up to several thousand parts can be casted, depending on their complexity, so that the molds are suitable for preseries but in principle also for series kickoffs. Figure 4-8 shows a model generted by DTM in an indirect metal sintering process that is inserted into a multipart injection molding tool as a core slide.

The quality of the surfaces has been improved again; there remain the disadvantages of the fragility of the green part, the risk of warping, and the overall extended processing time. Depending on the geometry of the parts, the build time could be as long as 20 to 30h, the post processing > 40h and the finishing 10 to 40 h bringing the total time up to 100h. This figure is, however, significant only if compared with the times needed for alternative processes.

Figure 4-8   Core slide produced with the DTM/3D-Systems sintering process. (Source: [BRE95])

The advantage of the RapidSteel process lies in the fact that the parts are entirely made of metal, even if it is only a kind of artificial alloy, and that they are therefore suited for plastic injection molding as well as for metal die casting. The process is in principle suitable for thixomolding.

With this process (quasi) series-identical functional prototypes can be produced. (see Example Supercharger, see Section 5.2.1).

Direct shell production casting (DSPC, Section 3.3.5.3), a 3D printing process, is also a member of this group. In this process the ceramic mold is generated by the injecting adhesive into a ceramic powder bed. The mold is then resintered and may be used directly for precision casting. The castings may be further processed to tools or tool inserts. The process is not widely known, as the marketing of the machine was stopped and the process is used only by external suppliers in the United States.

The multiphase jet solidification process (MJS, Section 3.3.4.2) is used for the production of models, prototypes, and tools. Apart from the single- and multiple-component plastic granules, powders and pastes are usually used today; metal and ceramic powders filled with polymers may also be used. This facilitates the production of tool inserts with properties similar to those produced in the metal injection molding (MIM) process. The ExtrudeHone 3D printing process (Section 3.3.5.2.) is a two step Rapid Tooling process which can be taken not only for metall functional parts but for dies and molds as well.

4.4 Metal Tools Based on Metal Rapid Prototyping Processes

Even though tools made from polymer-coated metal powders have proved to be reliable the aim is still to produce metal tools and mold inserts directly by sintering tool steel.

4.4.1 Multi-Component Metal Powder LASER Sintering

One step in this direction is a multicomponent powder technology in which a mixture of high and low melting point metal powders was developed in such a way that the low melting point component is melted in a classical plastic sinter machine and serves as "binder" for the high melting point component when it cools down and solidifies. The grain sizes and coefficients of expansion of the metals involved are matched in such a way that the resulting shrinkage is minimized as far as possible (see Section 3.3.2.3).

The EOS-Electrolux process uses basically a bronze-nickel powder mixed with a fluxing agent and other additives. The process known as direct metal laser Sintering (DMLS) is marketed under the name "DirectTool." To achieve the required accuracy the sintering has to be carried out carefully, resulting in a porosity of 30%. The stability is not high enough to enable these components to be used in injection molding processes. The components are therefore infiltrated with low melting point alloys of the "soldering" type or better still with epoxy resin. In this way the residual porosity is lowered to less than < 15%. The mechanical-technological properties are similar to those of aluminum. The inserts are said by some users to be mechanically workable by milling and polishing. The tools are suitable for use as injection molding inserts for preseries of several hundreds up to one thousand pieces. If the latest developments of metal powders such as Direct Metal 50V2 and Direct Metal 100V3 are applied, relatively soft materials with low abrasion produce the best results.

One of the advantages of producing molds by rapid prototyping processes is that 3D cooling ducts, deep grooves, or 3D-free form surfaces as well as comparatively filigree structures of down to 1 mm wall thickness may be realized. A new, even finer powder will allow the reproduction of details of up to 0.5 mm.

Figure 4-9 shows a tool with four sliders. All single parts were produced by the DMLS process [EOS98]. The components of the mold have been fitted by wire EDM. Specified literature claims that the build time can be reduced by 50% to 80% compared with traditional tool-making processes. These figures also are valid only for single cases and have to be checked thoroughly for each process.

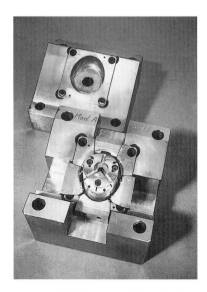

Figure 4-9 Direct metal process, tools. EOS-Elektrolux DirectTool. (Source: EOS)

4.4.2   Single-Component Metal Powder Laser Sintering

Direct single-component sintering processes using tool steel and other high melting point materials are being developed all over the world to avoid the limitations caused by the procedure and materials in plastic or multicomponent processes.

It is not only the higher sintering temperature attributable to higher melting points that is difficult to handle, but also some of the physical effects that cause the components to have a higher porosity than those produced in comparable plastic sintering processes. The higher surface tension of liquid metals causes large drops to be formed that result in a very porous structure after solidification. In addition, a layer of oxide builds up that lowers the wettability. The tensions that are thermally induced create warpings. It is technically hardly feasible to heat the entire powder bed up to some degrees below melting temperature – analogous to plastic sintering processes – if temperatures of between 800 °C and 1000 °C are involved.

On the basis of machine technology developed for the multicomponent process, EOSINT M250Xtended and the steel powder DirectSteel 50V1 were introduced in late 1998. The material is a specially developed steel alloy that possesses properties similar to those of steel castings after processing. The EOS DirectTool process thereby gained the status of a single-component process and problems with infiltration have been eliminated. The aim of enabling the direct production of inserts made of tool steel is thereby very nearly but not entirely reached. Reports show that important series materials such as polypropylene, polyethylene, ABS, and polyamide may be used.

Figure 4-10 SLPR process: tool for a lamp base, complete injection molded part. (Source: FhG-ILT)

Components of stainless steel with near 100% density are produced by means of the selective laser powder remelting process (SLPR) (Section 7.2.2). Internal tensions still cause deformations and the models are therefore built onto solid mounting plates. Optimized process parameters as well as improved scan strategies and optimized protective gas control help to conquer the problems. Using this process, complete two-part molds can already be produced from which injection molding parts can be made. Figure 4-10 shows a tool and the injection-molded lamp base that was produced with it after its completion. A disadvantage is still the rough surface.

The LENS process is based on the same principle (Section 3.3.6.1). No examples have yet been published of the production and testing of a complete injection molded tool.

4.4.3 Laser Generation

During the 1970s already thin-walled, shell-like components were produced by build-on welding for highly stressed nuclear power stations. This process was called "shape-welding." After it became possible with the aid of lasers to eliminate process obstacles, which seemed insurmountable up until then, and in view of the growing importance of rapid prototyping processes, the Americans took up the process and developed it further again as "shape welding."

Simultaneously the laser laminate process was developed further which led to the ability to control multi layer laminates with higher layer thicknesses. From there it was only a small step to the ability to controll the generation of freeform surfaces. The laser generation of arbitrary functional models and tools of any geometry is still a subject of research (see Section 7.2.3).

Two recently commercialized processes have emerged from the general approaches:

- **Generation with wire** (CMB, Controlled Metal Buildup, Section 3.3.7.2): The CMB process developed at the FhG-IPT, Fraunhofer-Institut für Produkionstechnologie in Aachen (Fraunhofer Institute of Production Technology, Aachen, Germany), with and for the milling machine manufacturer Röders enables the construction of layered 3D parts in a common high-speed cutter (HSC) with the aid of a diode laser and a wire feeding unit. By using the laser coating and the milling head in the same clamping and by changing from one to the other components may be produced simply by applying and removing. This development is primarily important for the repair of injection molds and also for the direct production of tool inserts.

- **Generation with Powders:** By optimizing the shielding gas control, the preheating, and scan strategies, the classical problems of surface qualities, warpings, and density are increasingly better controlled. The special advantage in comparison to buildup welding with wire lies in the wide range of available alloys achievable by mixtures of powders and, in comparison to powder bed sintering processes, in the simple construction of the machines. One machine is commercially available (Arnold, Ravensburg, Germany) that is basically a traditionally designed three- or five-axis milling machine and that is fitted with a special coating head, a 2-kW solid-state laser, and a commercial powder feeder. The nozzle geometry of the coating head is matched to the laser control in such a way that defined thin layers and wall thicknesses of a few millimeters can be generated. This machine configuration was also primarily developed to enable repairs to the near net-shaped form, for example, of turbine buckets. For this type of use the comparatively rough surface is no longer a disadvantage but a clear advantage compared with classical manual buildup welding. For use in the production of injection molding tools it will be especially important to develop a strategy analogous to the CMB process for the finishing of surfaces.

4.5 Summary and Perspectives

The suitability of all these processes has to be thoroughly checked in each case. When comparing processes with one another it is important not to compare the length of time needed for each single step but the length of time needed from the finished data set to the useable tool and, when comparing the budgets, to refer to those figures.

The aim of sintering tools and tool inserts directly from tool steel in commercial machines has still not been entirely achieved. Direct sintering in a powder bed is advancing tremendously but can still not be considered as commercialized as yet. On the other hand, wire and powder welding processes derived from laser welding processes have found their way to commercialization surprisingly quickly. Both variants are each combined with the machine concepts of a specialized producer and tend to be designed for the near net-shaped mold repair. Its potential for direct use in the production of injection molding and die casting tools is obvious.

Two examples for the use of rapid tooling are described in the Sections 5.2.2 and 5.2.3.

The interested user should follow developments closely. At present worldwide a great deal of effort is being put into the further development of indirect and direct tooling processes. If the development continues at this rate, processes for the direct sintering of tool steel will be industrially available in the near future. They will not replace tool making, just as rapid prototyping processes will not replace model making. They will complement it and make it faster, better, and in the end more effective.

5 Applications

The use of the prototypers described in Chapter 3 that are suitable for the optimization of product development are illustrated in this chapter by means of specially chosen examples. The examples concentrate on highlighting the advantages that can be achieved by employing rapid prototyping. The processes themselves remain more or less in the background. It cannot be assumed therefore that one process would in general be better suited than the others based on the type of process mentioned or the number of times it is mentioned in the examples.

At first, the examples concentrate on industrial product development. Thereafter, examples of rapid prototyping processes for experimental checks of calculating methods are described. The application possibilities are supplemented further by examples of rapid prototyping processes in art and design as well as in medical use, archaeology, history of art, and architecture and are intended to encourage the user to look for possibilities beyond daily routines.

Build times and costs are deliberately not mentioned because a serious discussion of that information is not possible on the basis of the information available and is beyond the scope of this book.

5.1 Rapid Prototyping in Industrial Product Development

5.1.1 Example: Pump Housing

Rapid prototyping task	Development of a prototype for a double-pump housing feeding two hot-water circulation systems from one thermal spring
Rapid prototyping process	Stereolithography, laser sintering
Model class	Geometrical prototype, functional prototype
Production process (series)	Injection molding with melting core technique

To facilitate the feeding of two separate heating circulation systems by one central thermal spring, a double-pump housing is necessary that can be adapted with little effort to various thermal springs on the one hand and on the other allows the accommodation of mass production circulation pumps.

The part was designed according to the general producer instructions with a 3D CAD model (IDEAS).

A 3D-visualizing software enables all concerned to observe the first construction step and to give their approval.

STL data were generated from 3D CAD data and a stereolithographic model (Fig. 5-1, right) was produced. With the help of this model the construction data were substantiated. Geometric and also casting technical assessments are necessary. Owing to higher mechanical and thermal strain a nylon model was sintered for the flow studies (Fig. 5-1, left).

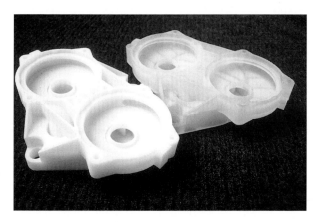

Figure 5-1 Double-pump housing. Stereolithography model (right), sinter model (nylon, left). (Source: M-Tec)

Because of the quick and reliable assessment by a 3D visualization and the stereolithography model, the production of prototype tools was not necessary. Therefore, the series tool was produced immediately. A special feature of the injection molding technique is the use of melting cores. This technology enables the production of a one-piece pump housing and avoids potential leakages as well as "seams" due to a mold-parting line that could be obstacles for the flow.

Figure 5-2 shows the injection molding part before the meltout of the cores and a separate core such as the one inserted into the injection molding tool.

The cores themselves, which are made of low melting point alloys, have to be produced in a tool. Here too, rapid prototyping processes offer an alternative to traditional eroding and milling processes. The melting cores can be produced, for example, by a tool molded of high-temperature silicone or in analogy to Example 5.2.4 according to the LOM process.

Figure 5-2 Injection molding part with melting cores, melting core (front). (Source: IKM, Essen, Germany)

5.1.2 Example: Office Lamp

Rapid prototyping task	Design and development of an office lamp product line
Rapid prototyping process	Stereolithography, vacuum casting
Model class	Geometrical prototype, functional prototype
Production process (series)	Plastic injection molding

A producer of lamps for offices and shops asked his design and production department to develop a new line of lamps. Prior to this, painstaking market researches had been made. The decisive impulse for the development of a new line of lamps therefore came from the marketing department.

The development target was, apart from the customary light-technical, electrotechnical, and production-technical requirements, to produce a distinctive but still rather conservative design that would integrate itself into business surroundings. In addition to various pendant versions that could be combined, a wall version and a floor version were required. A further requirement was the technically optimized production of this line of lamps.

In view of the restrictive pricing requirements for the final product, a feasible solution was quickly found. It was based on a simple basic concept using an economical production technique in which variations could be realized by individualizing the basic concept.

Based on this idea by the industrial designer concept (TRENO, Finland), which stems from the renowned German Rail Zeppelin concept (Fig. 5-3), a four-part basic concept was developed consisting of three different elements. An elongated lamp body, which is already in the prototype phase and is made from a continuous casting aluminum profile, forms the actual lamp together with two cap elements at the head ends. A suitable mounting element allows the alternative design for a floor lamp or a pendant lamp. The subsequent drafts assist in forming a consensus concerning the basic design of the central part and the head ends of the basic lamp. Based on the design drawing a support element for the floor version was favored (Fig. 5-4, left).

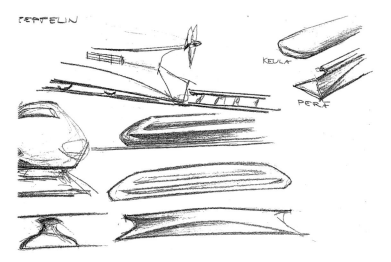

Figure 5-3 Design draft "Rail Zeppelin." (Sketch: TRENO)

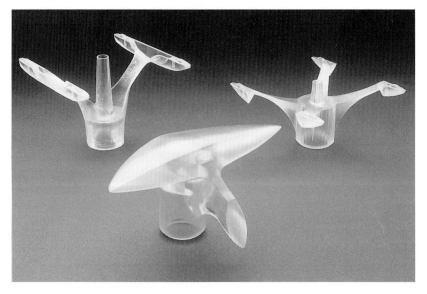

Figure 5-4 Support elements, floor version. (Source: THORN)

To promote further product development it was decided to produce prototypes for the whole lamp, and especially of the head ends and the support elements. As the company had its own conventional model making department it would have seemed understandable to employ a manual or semiautomatic conventional production. For time reasons, however, it was decided to use the then new technology of rapid prototyping specifically laser stereolithography. The fact that the extremely intricate free-formed area of the draft would have been very difficult to produce by conventional means made the decision for rapid prototyping technology easier. Here, the laser stereolithography especially promised a fast and geometrically exact solution. The cost of making the models was taken into consideration only insofar as it was not allowed to be much higher than that of a conventional production. The cost had to include the 3D CAD representation which was not necessary for conventional production. The crucial factors for the decision were, however, the complexity of geometry, the speed of model making, and the possibility to test this promising new technology extensively while bearing the company's own product requirements in mind.

During the first project phase already it was decided not to produce the entire lamp as a stereolithography model but to produce each of the main elements by the specially suited process and to assemble the lamp from these parts. The central element of the lamp was made of an continuous casting aluminum profile, as already mentioned, while the two head elements and the support element were made as stereolithography parts. The stereolithography parts were meant to be used as a sample. For further product development it was decided to use them as master models for vacuum casts. The vacuum casting became necessary owing to the number of pieces required but mainly because the stereolithography material showed a different translucency than the series material.

The head pieces immediately met with general approval and were altered only insignificantly later. The stereolithography model and the corresponding vacuum cast are shown in Fig. 5-5.

In contrast, the authentic scale production of the mounting elements (Fig. 5-4, left) showed that this element was too filigree in comparison to the entire construction and optically too unbalanced. Although the second version of the mounting element met with approval when presented as an alternative draft (Fig. 5-4, right), it was later discarded after having been produced and duly assessed as a 3D stereolithography model.

Only the third variant (Fig. 5-4, middle) met with general approval and this was taken into series production. A final judgment was first possible after this element, which was difficult to represent in a drawing, was actually presented as a physical model. In such a complex design dominated by free-form areas, which also had to accommodate the internal cable, the advantages of rapid prototyping processes became obvious and, in this specific case, the ability of stereolithography to depict free-form areas very accurately.

All in all, the implementation of stereolithography enabled an earlier and more reliable appraisal of the product and its basic elements and thereby a reduction in product development time. The multinational dialogue which, in international companies is at times inhibited, was also positively influenced. The product could also be presented in its final appearance at important trade fairs in the form of stereolithography models and their subsequent vacuum casts. Figure 5-6 shows the series product (head piece). The pendant version especially is clearly reminiscent of the German "Rail Zeppelin" (Fig. 5-7).

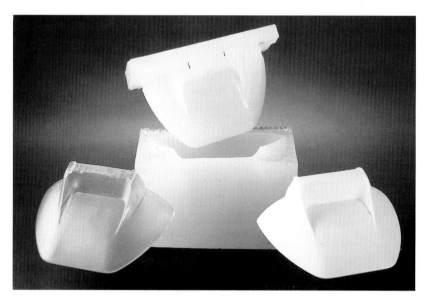

Figure 5-5   Head pieces. Stereolithography model (left), vacuum casting mold (middle), vacuum casting (right). (Source: THORN)

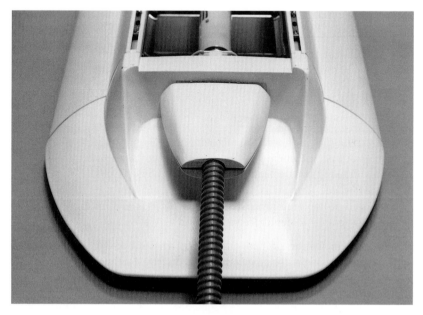

Figure 5-6   Series lamp, head piece. (Source: THORN)

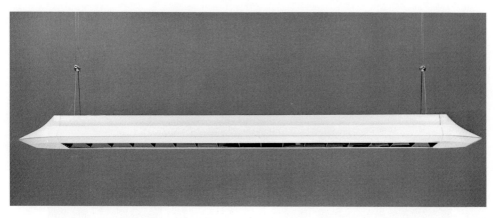

Figure 5-7 Series lamp. (Source: THORN)

5.1.3   Example: Integrated Lamp Socket

Rapid prototyping task	Development of a lamp socket
Rapid prototyping process	Selective laser sintering
Model class	Functional prototype
Production process (series)	Plastic injection molding

While developing a three-piece casing for a plug-in halogen luminous element a functional prototype was required that enabled the accuracy of the fit of the multiinsert socket, adaption ring, and socket case to be monitored. As required of the series product, the functional model had to be relatively stable under high temperature to enable the functioning of protruding brackets and fine guidances especially to be tested. The actual surface finish was of secondary importance as long as it did not impair the function. Therefore it was decided to sinter this part of nylon. Figure 5-8 shows the sintered piece and a series piece that was checked against the sintered piece and production technically derived from the sintered piece. It is produced in series as injection molding piece. The rapid prototyping model allowed the piece to be optimized by using original luminous elements before an injection molding die was made.

In Section 4.4 the production of the corresponding injection mold by means of rapid tooling is described (Fig. 4-10).

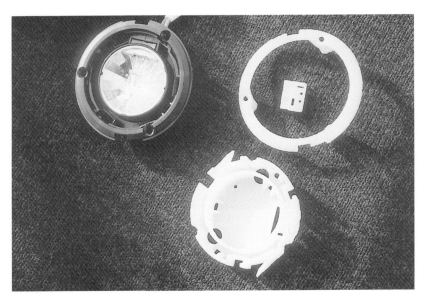

Figure 5-8 Lamp socket. Sintered prototype parts (white), Series part (black). (Source: CP)

5.1.4   Example: Model Digger Arm

Rapid prototyping task	Production of a model digger arm
Rapid prototyping process	Stereolithography Selective laser sintering
Model class	Geometrical prototype, functional prototype
Production process (series)	Die casting

To clean and empty hot-metal ladles, pouring ladles, and converters of heavy slag conglom-erates ("skulls") special digger arms fitted with a ripper tooth are mounted onto standard digger undercarriages. The basic construction and the functioning of the special digger arm are shown in Fig. 5-9.

The producer of these special digger arms required a model to a scale of 1:50 for advertising purposes and also a fully functional model with remote control to a scale of 1:20 for trade fairs and exhibitions.

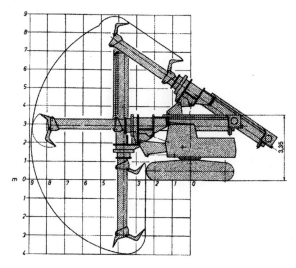

Figure 5-9 Special digger arm, kinematics, and work area

Since authentic scale models of diggers are available in shops, the development focused on the production of matching digger arm models. The smaller arm was generated with the aid of stereolithography because this demanded detailed accuracy while the large arm was sintered because this demanded mechanical resilience.

The main challenge lay in reducing the scale which only superficially resembles the reduction of all geometrical measurements by a certain factor. In transcribing the characteristics of the original to the model, details such as screwed connections need to be significantly larger in scale than in a linear reduction. The aim of using a model for design is to discern a high degree of detail and to be able to perform all important kinematic functions of the original digger arm. This demands minimum requirements concerning functional areas and wall thicknesses which differ from the geometrical reduction and need to be correlated with the above-mentioned optical requirements. The requirements on the model grow to the extent to which assembly and production requirements for a subsequent production of small series – from several hundred up to one thousand pieces – in metal or plastic are further integrated into the model.

As the original construction was not based on a 3D drawing, the model making followed a 3D CAD model (ProEngineer). Figure 5-10 shows the 3D CAD model in the assembled form and exploded view of the two main elements of the digger arm – the external structural components and the internal structural components of the digger arm. The two exploded views together show every single part which is later generated by rapid prototyping processes.

Figure 5-11 shows the ripper tooth as a stereolithography model, Fig. 5-12 as a sinter model.

Figure 5-13 shows the finished stereolithography model in the assembled form and after having been mounted onto a typical toy digger undercarriage to scales 1:20 and 1:50.

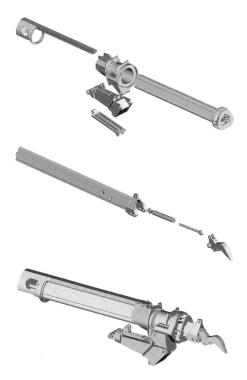

Figure 5-10 Special digger arm. External and internal structural components, assembly drawing (seen from above). (Source: van Crüchten/FH Aachen, Germany)

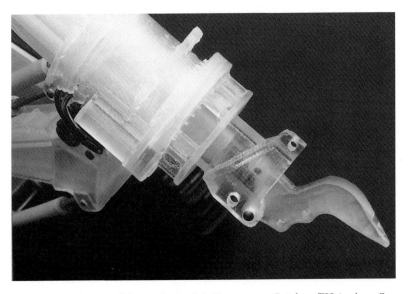

Figure 5-11 Ripper tooth, stereolithography model. (Source: van Crüchten/FH Aachen, Germany)

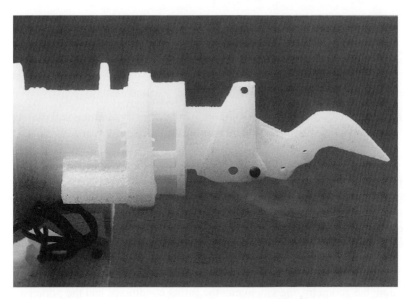

Figure 5-12 Ripper tooth, sinter model (nylon). (Source: van Crüchten/FH Aachen, Germany)

Figure 5-13 Digger model to scales 1:20 and 1:50. (Source: van Crüchten/FH Aachen, Germany)

Rapid prototyping processes allow locally different scales to be optimized and thus achieve the overall optical impression of series quality. The desired miniseries is not feasible economically owing to tooling costs. This perspective is opened up by combining rapid prototyping processes with molding processes.

5.1.5 Example: LCD Projector

Rapid prototyping task	Geometrical and functional testing of an LCD projector design
Rapid prototyping process	Stereolithography, vacuum casting
Model class	Stereolithography: geometrical/functional prototype Vacuum casting: technical prototype
Production process (series)	Plastic injection molding

LCD projectors must become ever more compact while their light transmitting capacity must increase. In addition, the demand for a functional, up-to-date, and representative design is growing. These development targets sometimes work in opposite directions and an intensive coordination between designers and technical engineers is necessary. The model shown has a very compact design inasmuch as it can be folded for transport.

This reveals two clear demands on the model maker:

• The design, the geometry, the kinematic, the fitting conditions, and the principal function must be monitored.

• The thermal resilience of the casing must be monitored.

The visualization of the design, its external form, the kinematic, and thereby the principal function make great demands on the quality of the surface, the reproduction of details, and the fitting qualities of the model parts. Thermal and mechanical strains are quite slight in this phase. For this reason a multi piece stereolithography model was built initially (Fig. 5-14).

It is not possible to monitor temperature changes, which also involves checking the necessary cooling systems for continuous running, with a stereolithography model since continuous running temperatures of over 60 °C already pose a problem. The part, therefore, was molded by vacuum casting (Fig. 5-14). The required temperature stability of 130 °C, however, creates difficulties with most pourable resins. For this reason, in this case, a suitable temperature-stable pourable resin was used.

A further advantage of vacuum casting is that in this way several casings can be produced in spite of the large and complicated contour, owing to many fine cooling ribs. At the same time, the castings aided the production of series injection molds that was taking place parallel to the prototyping.

The extremely compact construction of the device together with the mechanism that puts the lens into projection position when the casing is opened is a great design challenge both from the mechanical point of view and – because of the high temperature – thermally. For such complex flow patterns no reliable theoretical calculating methods exist as yet. The necessary experimental monitoring of the design was made possible in a fast and reliable way with the aid of a series-like prototype generated by stereolithography and vacuum casting long before a series tool was produced.

Figure 5-14 LCD projector casing: stereolithography model (above) and vacuum casting (below). (Source: CP)

5.1.6   Example: Capillary Bottom for Flower Pots

Rapid prototyping task	Production of a two-piece capillary bottom with film hinge for watering flower pots automatically
Rapid prototyping process	Selective laser sintering
Model class	Technical prototype
Production process (series)	Plastic injection mold

Helped by a prototype, a plastic tray is to be tested that waters flower pots automatically by using capillary forces. The plastic tray used for larger flower pots is nearly square and has an integrated folding reservoir that is connected to the main part by a film hinge.

It was decided to produce prototypes by means of selective laser sintering and to test the functioning of the film hinge and other mechanical-technological properties; the functioning of the parts produced by the injection molds under authentic usage conditions was simulated (Fig. 5-15). The fact that the polyamide used is hygroscopic and that the part would therefore change to a small extent when used was not considered a disadvantage for its intended purpose in these tests as the functionality was not impeded by such tiny tolerances.

Directly after completion of the 3D design rapid prototyping made it possible to test all relevant characteristics of a functional model under the same conditions as those for which the series parts would be used. This was the basis on which the design was optimized before tools were produced.

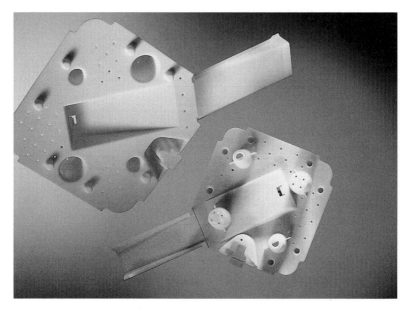

Figure 5-15 Capillary bottom for flower pots produced by selective laser sintering. (Source: CP)

5.1.7 Example: Casing for a Coffeemaker

Rapid prototyping task	Production of complete coffeemakers as models for trade fairs that are externally identical to the series product
Rapid prototyping process	Stereolithography, vacuum casting
Model class	Geometrical prototype, functional prototype
Production process (series)	Plastic injection molding

The models had to be produced in various colors, the parts had to fit perfectly, and high demands were made on the quality of the surfaces. Therefore, stereolithography was chosen as a rapid prototyping process. To realize various colors and to be able to produce several prototypes the stereolithography master models were molded by vacuum casting. In this way it was possible to use various plastics, sometimes transparent, for water transmitting internal parts, the casing, and the water level gauge while still using the same master models and molds.

With rapid prototyping it was possible to cast casings in various colors, some of them even transparent, without having to produce tools at this very early stage of product generation. The use of series parts such as glass jugs, turning knobs, lamps, and so forth gave the impression of this model being a series product long before the decision to proceed with series production was made.

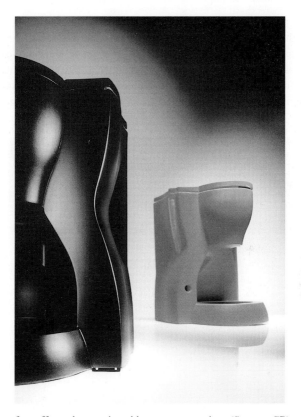

Figure 5-16 Casing of a coffeemaker produced by vacuum casting. (Source: CP)

5.1.8   Example: Intake Manifold of a Four-Cylinder Engine

Rapid prototyping task	Production of an intake manifold for a four-cylinder engine
Rapid prototyping process	Selective laser sintering
Product Development Stage	Testing
Model class	Functional prototype/technical prototype
Production process (series)	Plastic injection molding /metal die casting

Intake manifolds for internal combustion engines are becoming increasingly more complex because adjustable elements are being built in to ensure the continuous adjustement of the flow conditions to the engine characteristics. In addition, a number of sensors and actuators are built into the intake manifold. Aerodynamic tests are inadequate even with numerical 3D calculating methods primarily because of the unsteady flow. After a theoretical preoptimization, therefore, it is still indispensable to run a test on the engine test bench.

Mounted on the vibrating engine, the demands made on the mechanical and to a lesser degree on the thermal resilience of the prototype are very high; the active baffles must function and the sensors and actuators must be introduced. For these reasons, therefore, a sinter component made of polyamide is usually chosen especially in cases where the later series component is produced in plastic injection molding. For the intake manifold shown (Fig. 5-17), an FDM component made of ABS plastic would also have been a good alternative.

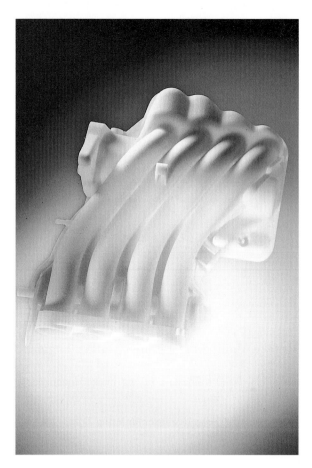

Figure 5-17 Intake manifold of a four-cylinder engine made by selective laser-sintering. (Source: CP)

With rapid prototyping a useable functional model is available very shortly after the 3D design is finished. Geometrical alterations, which may become necessary after the test run on the engine test bench, can be effected in an optimized prototype comparatively fast and economically. Another advantage is that alterations to the theoretical calculation program and the resulting practical consequences can be correlated directly on the engine test bench and are thus the basis for an optimization strategy.

5.2 Rapid Tooling in Industrial Product Development

5.2.1 Example: Turbocharger Casing

Rapid prototyping task	Production of a turbine housing for an exhaust gas turbo-charger for hot gas tests on a test bench
Rapid prototyping process	Laser sintering
Model class	Technical prototype
Production process (series)	Steel casting

The intake spirals of turbine housings for exhaust gas turbochargers have a decisive influence on the operating performance of the engine and thereby on its efficiency, torque, consumption, and exhaust emission. Turbochargers with two or more spirals especially can be designed in such a way that optimal conditions are also achieved when operated at lower speed.

It is still not possible to make a complete and accurate calculation of the extreme 3D flows in the narrow spirals. In the last instance therefore bench tests need to be carried out to verify theoretical reflections. The time-consuming production process, which is based on conventional sand casting processes and takes 2 to 3 months, means that optimizations based on test bench tests have been extremely time consuming until now.

By sintering steel powder directly the production time can be reduced to 1 week (see Section 3.3.2.2).

Figure 5-18 shows the sintered part still with the supports attached before its mechanical finishing. This can easily be accomplished by traditional milling processes such as drilling, machining, turning, and polishing.

Figure 5-19 shows the completely finished housing ready to be fitted. Figure 5-20 shows a similar cutaway housing from series production.

The turbine housing was completed with a compressor and turbine unit and tested on a hot-gas test bench at 570 °C (exhaust gas temperature of diesel engines). The user was very satisfied with the result and also with the surfaces. Although they showed a stair-stepping effect, this was judged to be not worse than the well known surface roughness of sand castings.

Figure 5-18 Metal sintered housing for a turbocharger (RapidSteel). (Source: 3D Systems / DTM)

Figure 5-19 Metal sintered housing for a turbocharger, ready to be fitted. (Source: 3D Systems / DTM)

Figure 5-20 Cutaway model of a series turbocharger housing. (Source: 3K-Warner)

By using this method the fast and individual adjustment of exhaust turbochargers to engines is possible to better effect and faster than previously.

The process is even more effective if the geometry is verified by a plastic model before the metal model is produced.

5.2.2 Example: Two-Piece Electric Connector

Rapid prototyping task	Production of a small series of electric connectors
Rapid prototyping process	Stereolithography, silicone molding process, casting in low melting point metal (LMPM casting)
Model class	Technical prototype
Production process (series)	Plastic injection molding

The producers of electric connectors need prototypes at a very early stage that already possess the fundamental properties of the subsequent series product.

Prototypes made of stereolithography material, especially if used in motor vehicles, are usually unsuitable because of the operating environmental conditions (temperature, humidity, mechanical strain). Prototypes made of polyamide (laser sintering) or ABS (FDM) are in most cases able to accommodate themselves to the conditions in the operating environment but very often they do not fulfil the requirements concerning the high-quality detail reproduction demanded today of multipolar connectors with compact dimensions.

One solution could be a combination of stereolithography for high-quality detail reproduction, and surfaces and molding processes that allow the production of connectors from materials possessing the necessary mechanical-technical properties. Figure 5-21 shows the stereolithography model of one side of the connector.

When casting the connector in a mold, a set of three cores is required for each side (Fig. 5-22). Theoretically, this could also be produced by stereolithography as illustrated, but the cores show such fine, protruding details that it is not possible to draw them. Such cores therefore are milled preferably from metal, or eroded.

Figure 5-21 Stereolithography model of an electric connector. (Source: FCI/CP)

As an alternative to milling the cores, a process chain was developed using metal cores that are made, as the later connector modules, by filling in silicone molds.

This requires the following process steps:

- The cores are derived from the CAD data set for the two-piece connector components which are produced together with them as stereolithography components. Figure 5-22 shows the core set for one side of the connector.

- In addition, a stereolithography model of the entire external contour of the connector with the three inserted cores is made.

- The cores and the complete model are molded separately in silicone.

- The mold for the core is then filled in with low melting point metal (LMPM, see also the casting procedure when using Keltool which is in principle identical, (Fig. 4-4)) and the metal cores are taken out after solidification. Figure 5-23 shows a cast and the corresponding stereolithography model.

- The cores are then inserted into the corresponding recesses of the external contour model and the connector is cast around this by a vacuum process. Figure 5-24 shows the connector with the cores after it is removed from the mold, but before it is melted out.

- The metal cores cannot be drawn out of the cast case because they were cast directly from the stereolithography part and the fine details and their surfaces were not finished. As they are made of low melting point metal they can be melted out at a low temperature.

Figure 5-22   Three-piece stereolithography model of a core set of an electric connector for the component in Fig. 5-21. (Source: FCI/CP)

When the silicone mold is filled several times it is possible to produce a small to medium series of cores that are used later as lost cores when vacuum casting the casing. The process described in the preceding must be carried out for each part of the connector so that in all six cores and two casting operations per connector unit are necessary.

The generative process chain is useful if there is any danger that milled cores could be damaged during the drawing. This can easily happen owing to filigree structures. If there is no such danger then the milling method is neither slower nor cheaper, but it does produce surfaces of higher quality. The generative method is preferable if a large number of very filigree pins with narrow deep slits has to be realized, or if, because of to a defect, several cores have to be produced. The time lost on repeated millings is a negative factor even more so than the additional cost of milling a second set of cores. A second, third, or fourth set of cores is quickly cast with the generative process. This process therefore enables the speedy and economic production of small series and the parallel casting of several connectors simultaneously. The milling process would require a second, third, and fourth set of cores each of which would cost the same and would take the same time to produce as the first set.

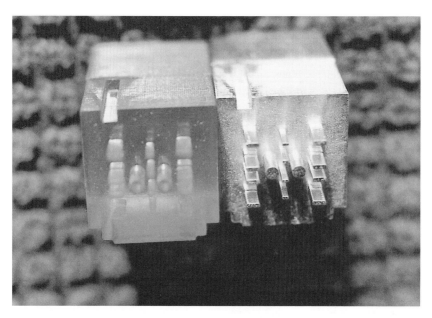

Figure 5-23 Stereolithography core and cast of an electric connector produced by the LMPM process. (Source: FCI/CP)

Figure 5-24 Electric connector with cores of low melting metal. (Source: FCI/CP)

5.2.3   Example: Injection Mold Headlight Dipper Arm

Rapid prototyping task	Production of a two-piece metal injection mold by direct sintering of a two-component metal powder
Rapid prototyping process	Selective laser sintering (EOS)
Model class	Technical prototype, series part
Production process (series)	Plastic injection molding

To check whether specific headlight dipper arms could be produced and to check their functioning an injection mold should be made by sintering metal powder directly from which several hundred prototype pieces can be produced.

The part produced in the injection mold is an automobile headlight dip switch arm measuring approx. 55 · 32 · 20 mm. This part has freestanding ridges of width 1.5 to 2 mm which are to be shown as corresponding slots in the model.

The construction data for a two-piece injection mold were generated from a bronze alloy using an EOSINT M laser sintering machine. The build time was 9h. The infiltration with tin took about 1h. Afterwards, the part was polished manually, an indispensable process even when milling techniques are employed. The surfaces were finished by eroding (EDM). It is interesting to note that the standard parameters for the EDM processing of copper-steel proved to be very suitable. The sprues and spreaders were generated by milling. The tool was subsequently tested using various materials in the injection molding machine. A total of 300 pieces was produced from PE (polyethylene, 180 °C), PP (polypropylene, 210 °C), and ABS (250 °C). The cycle time was on average 40s. The nozzle pressure was 600 bar. Even after further tests with glass-filled nylon (PA 6/6 with 30% glass, 260 °C) the first slight rounding was noticeable only at the sharp corners of the mold. This mold could undoubtedly produce another few hundred pieces.

The tests were made in a typical injection molding machine with typical injection molding material. The tool itself could also be sintered from a single-component steel powder. The stability of the mold could be expected to be higher, but the finishing work for the harder material would require more effort.

Figure 5-25 shows one half of the mold and the injection molded part from various viewpoints.

Figure 5-25 Directly sintered injection mold, injection molded pieces. (Source: EOS)

5.2.4   Example: Permanent Mold for Precision Casting Models

Rapid prototyping task	Permanent mold for the production of lost wax models for the precision casting of golf-club heads
Rapid prototyping process	Laminated Object Manufacturing (LOM)
Model class	Technical prototype
Production process (series)	Precision casting

When structuring an injection molding process for the economical production of heads for golf clubs, a separable mold has to be produced with which the necessary lost wax models are generated.

On the basis of the desired positive, a correspondingly scaled two-piece negative mold was designed as an LOM model. After the surface has been properly finished the mold is used for producing the lost wax models needed for the precision casting process. The separable mold is therefore treated with release agent. The molds are very resilient if treated carefully and enable the production of dozens of wax master models (in the example 50) without noticeable loss of quality.

If desired, other materials can also be cast in the same manner. Figure 5-26 shows the two-piece LOM model with added fitting drillings and a wax model inserted for demonstration. In the lower part there are three precision casted unmachined blanks and two finished golf club heads.

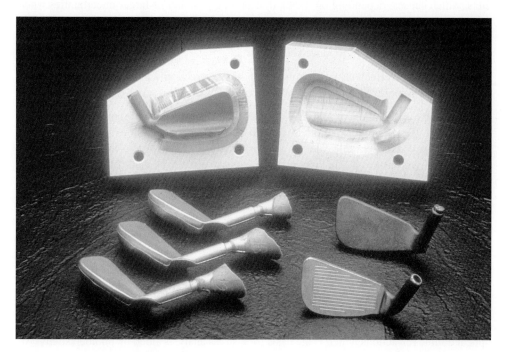

Figure 5-26 LOM mold for the production of wax models as lost models for precision castings. Precision casting blanks, finished golf club heads. (Photo: Cubic Technologies / Helisys)

Design alterations are quickly implemented with the aid of rapid prototyping. New golf club concepts can even be tested as series identical models by selected customers before mass production is started. Individualized miniature series can be produced economically by this process.

5.3 Rapid Prototyping for the Evaluation of Calculating Methods

5.3.1 Models to Be Used in Calculating Methods – General

Mathematical-physical calculating methods such as computer models are a fast, effective, and cheap aid for the technical designer. They allow the simulation directly in the computer of product properties such as stability, vibration behavior, temperature resilience and of production methods such as filling simulations based on purely theoretical approaches. The application of such processes, which also form the basis for virtual reality methods, is the fastest way to create new products.

The disadvantage is that all calculating models have to work more or less on simplified facts, which causes, depending on the kind of product, smaller or larger uncertainties when the results are incorporated into the series production. Therefore, the variables within the models should be checked as often as possible and corrected, if necessary, according to appropriate tests made with produced series parts.

A faster and more effective way of obtaining such correction values could be an experimental check of appropriate rapid prototyping models instead of series components.

In practice the calculation of stability for new products is an important part of product development. The theoretical models contain such large uncertainties that tests on real components are indispensable. Here, rapid prototyping models come into their own and open up new possibilities. More precise statements can already be made on the model owing to a systematic analysis of analogies between the optical or thermal properties and the strain in the component. These processes are known as photoelastic or thermoelastic stress analysis.

Photoelastic Stress Analysis

The approach is to use the refraction properties of stereolithography models for photoelastic analyses and thereby verify the predictions of computer calculations by tests at an early stage of product development.

Epoxy resins, especially those used for stereolithography, have the property of displaying optical double refractions which is the basis for a stress analysis aided by photoelasticity. An interference pattern in the transparent stereolithography component is obtained that is proportional to the stress within the component and that can be interpreted manually or automatically.

This process adopts the concept of photoelasticity, which has been known since the beginning of this century. With the aid of stereolithography models today's greatest problem can also now be solved – that of making 3D models from conventional photoelastic materials.

Epoxy resins are exceedingly well suited for the generation of models for photoelasticity because of their high transparency. They hardly differ from the standard materials (Araldit B, Ciba Geigy) used for photoelasticity. Acrylates are less well suited because of their lower degree of transparency. If used for stereolithography models care should be taken that the model is cured as evenly as possible, that as few air bubbles as possible are created, and that the surface is flawless. Some kinds of constructions and exposure methods, for example, those generating hollows intentionally to increase the quality and dimensional stability of SL components, are therefore unsuitable for photoelastic checks. It also has to be taken into account that semi polymerizing processes (whereby complete polymerization occurs later in a post-curing oven thus create optically effective parting planes. From the aspect of photoelasticity, components made of epoxy resins SL5170 and SL5180 which use the build style ACES (3D Systems) are especially suitable.

The basic setup for photoelastic checks (Fig. 5-27) consists of a pole filter, a light source, and two λ/4 wave plates, which facilitate the separation of the isochromatics (lines of the same main stresses differences) from the isoclines (lines of the same main stresses) so that then only the isochromatics and the transparent model, which is placed between the two λ/4 plates can be viewed.

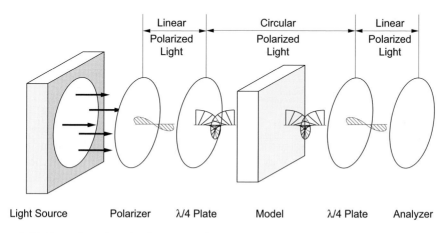

Linear	Circular	Linear
Polarized Light	Polarized Light	Polarized Light

Light Source   Polarizer   λ/4 Plate   Model   λ/4 Plate   Analyzer

Figure 5-27 Setup for a photoelastic stress analysis. (Drawing: Steinchen)

Regardless of the type of resin used, the necessary material parameters must first be defined before the photoelastic test can be carried out and evaluated. The most important value is the photoelastic constant which is determined from a calibration test and which denotes the relationship between stress and isochromatics order. The photoelastic material parameters of the most important stereolithography resins are listed in Appendix A2, Figure A2-15 [STE94]. Should other resins or other build styles (scan strategies) be used, these calibration tests ought to be repeated.

Thermoelastic Stress Analysis (THESA)

The development of safety relevant devices such as, for example, steering assembly parts in automobile manufacture, still relies on tests made on series identical models. Rapid proto-typing processes change little here since the plastic parts and also modern metal parts can either not be tested under series conditions, or – if tested – they do not allow any applicable conclusions to be drawn.

THESA is successfully employed in testing such components without extensive field trials. It allows single parts to be simulated on the test bench instead of monitoring the module or a driving test. THESA is based on the fact that metal components under stress show tempera-ture changes that are proportional to the given stress, provided one stays within the elastic sector. These temperature fields and their fluctuations can be recorded using appropriate high-resolution thermo cameras; tensions can be related to the temperature fields and correlated with the strains. This process was modified in cooperation with an automobile manufacturer so that it is now possible to use this process, initially meant for metal components, for plastic components also [GAR98]. In principle, this has paved the way for the optimization of highly stressed components with the aid of plastic rapid prototyping models.

5.3.2   Example: Photoelastic Stress Analysis for a Cam Rod of a Truck Engine

Rapid prototyping task	Production of a valve lifter model for checking the calculated tensions with the aid of photoelastic analysis
Rapid prototyping process	Stereolithography
Model class	Geometrical prototype
Production process	–

The starting point is a 3D CAD model of the component to be tested. From this a stereolithography component is built. Care should be taken that the value of the photoelastic constant is kept as low as possible. The most important parameter here is the postcure under UV light [STE94].

Further action depends on whether a 2D test at room temperature is to be run, or a 3D photoelastic test. If a 3D test is required the stereolithography component must first be "frozen." Using an oven with a programmable cooling curve, the model is heated up to approximately glass transition temperature, loaded, and then cooled down to room temperature at 2° to 3°/h. Afterwards, the elongations, and thereby also the tensions, are "frozen."

The photoelastic test itself is now run on 2D models. In the case of a 3D photoelastic test this involves the cutting of a cross section at the test-relevant area for use as a 2D model. A normal band saw can be used for this purpose.

When placing such a 2D (Fig. 5-28, top) or 3D (Fig. 5-28, middle) model of the valve lifter (Fig. 5-28, bottom) between two polarizors or quarter-wave plates (Fig. 5-27) a useable isochromate pattern is recognizable in the 2D model. For the 3D model of the valve lifter the isochromate images of the single layers are superposed so that a clear definition is not readily possible. In general, it can be observed that the stress concentration is higher the more isochromates appear at a particular point and the closer they are together. If the quarter-wave plates are removed the isoclines are recognizable. Whether an interpretation of the isocline picture is necessary depends on the method of analysis used. An automatic analysis using an electronic image processing system is also possible. The aim of the analysis is to define the main strain and the main stress to enable a conclusion to be drawn concerning the numerical value of the load.

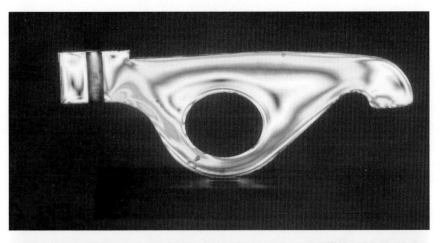

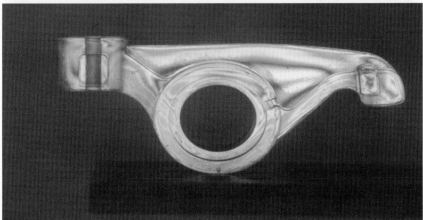

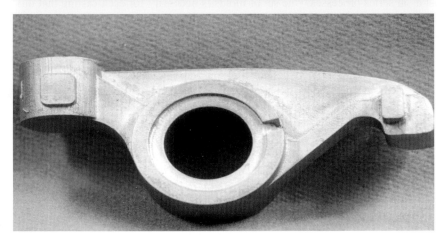

Figure 5-28 2D (top) and three-dimensional (center) model of a valve lifter (bottom). (Source: Steinchen)

After the interpretation of the isochromate picture is complete, the results of the test can be used for dimensioning prototypes with the aid of similarity laws. The similarity law for the static case is, for example:

$$\frac{\sigma_P}{\sigma_M} = \frac{F_P}{F_M} \cdot \left(\frac{l_M}{l_P}\right)^2 \tag{5-1}$$

As Equation (5-1) shows, the results are transferable without great effort. To summarize, it can be stated that photoelasticity together with stereolithography is a reliable and economic method of implementing an experimental optimization of the subsequent product already in the design phase.

5.3.3   Example: Thermoelastic Stress Analysis for Verifying the Stability of a Car Wheel Rim

Rapid prototyping task	Production of a segment of the rim of a car wheel for a thermoelastic tension analysis
Rapid prototyping process	Laser sintering
Model class	Geometrical prototype
Production process (series)	–

The wheel rims of sports cars are highly loaded safety components that are increasingly required to be lightweight and aesthetic while the dynamic stress on them is increasing continuously. A hollow rim promises to fulfil these requirements. The development is difficult, however, because the prototypes of hollow rim segments cannot be produced by milling as in solid constructions. For optimization purposes, therefore, the components must be cast as in series processes. One set of molds is necessary for each casting that has to be made as custom-built models from wood in a model workshop.

The development up to today has been correspondingly complicated (Fig. 5-29, left string). From the CAD data, which are checked and optimized by FEM processes (for reasons of clarity the optimization loop is not shown in the graph) a cast in aluminum is obtained by conventional mold making. This cast is finished, mounted, and tested. Depending on the results, this loop is performed several times and the results iteratively improved.

When THESA is employed, this process is shortened, as assembly and driving tests are omitted.

When using rapid protoyping processes the process can be further shortened insofar as the mold making is substituted by a rapid prototyping model that is cast directly in a precision casting process.

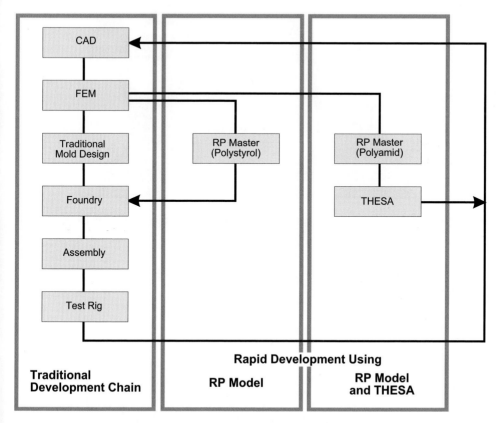

Figure 5-29   THESA. Schematic steps in the process chain and the shortening thereof by the use of
rapid prototyping. (Source Schwarz/CP)

The decisively shorter development process is achieved by being able to test the sintered
polyamide rapid prototyping component directly by means of THESA, thereby eliminating
the necessity of the entire casting process. Figure 5-30 shows an exemplary thermographic
image with a defined stress and, in comparison, the same situation with a molded aluminum
wheel rim. The conformities are excellent, as is documented by the totals of the main
tensions given in the example over the radius at the center of the component (Fig. 5-31).

When THESA is used for sintered plastic components generated by rapid prototyping, the
time needed for the iteration loop is reduced to 20% of the time previously needed. The
greatest single effect is the reduction of production time from 18 days for a cast component
to 3 days for a polyamide component. By applying this modified THESA process, the entire
development process can be reduced dramatically, although not down to 20% of the time
previously needed since the final test for a safety component will still be run with a
component made of the original material.

Figure 5-30 THESA. Thermographic reproduction of a rapid prototyping component (polyamide) (left) and of an aluminum series part (right). (Source: Schwarz/CP)

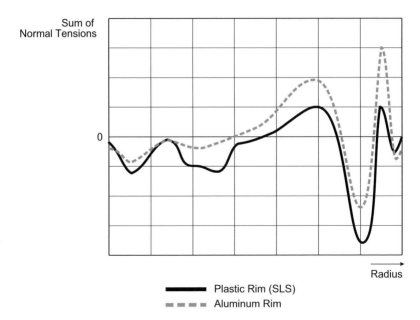

Figure 5-31 THESA. The total of the main tensions over the radius in the center of the component. (Source: Schwarz/CP)

5.4 Rapid Prototyping in Medicine

Rapid prototyping processes are used in medicine in two main areas: the visualization of invisible structures and the production of individual implants.

5.4.1 Anatomic Facsimile Models

3D physical models are used preferably for medical diagnosis as a supplement to 2D images. As these models are 3D images of the original they are called anatomic facsimile models (AFM).

The process chain starts with the medical indication calling for a model for diagnosis or therapy (Fig. 5-32). The data are obtained by means of classical processes. Computed tomography (CT) is used for reproducing bone structures, magnetic resonance imaging (MRI) and ultrasonic processes (US) for soft tissue, and positron emission tomography (PET) and single-photon emission computed tomography (SPECT) for the reproduction of circulatory and metabolic disorders. Although CT images, for example, also contain layer information, they cannot be used directly in the prototyper as the layers with 1 to 2 mm (minimum 0.5 mm) are significantly thicker than those used in rapid prototyping.

Using special image analysis processes, 3D images are reconstructed from the measured values. Each layer is segmented depending on its required structure. If bone and soft tissue is to be reproduced simultaneously, as is the case, for example with the larynx, single partial images may be superimposed. From the segmented layer information a virtual picture is created on the computer screen (Fig. 5-32) with the aid of 3D reconstruction processes. Modern CT scanners and image analysis devices are able not only to provide output media and data structures for the daily medical routine but also to produce STL data and to address prototypers directly. When this is not directly possible, special programs, for example, MIMICS (Materialice), are used. They also handle specific rapid prototyping problems (Section 3.1) such as positioning and orientation in the build chamber, any necessary supports, etc. This creates a raw-data basis that is equivalent to CAD models for mechanical engineering.

The rapid prototyping model can be used directly or as a master model. Soft tissue or details, highlighted in different colors, can then be reproduced more readily with the aid of molding processes, especially when using various materials. Depending on the definition of the problem it may be necessary to produce several models by molding processes.

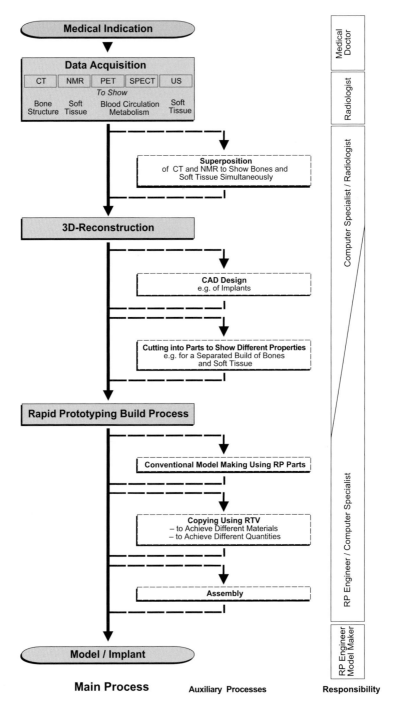

Figure 5-32 Process chain for the production of an anatomic facsimile model

5.4.2   Taylored Implants

In the course of a therapy, anatomic facsimile models are well suited as a basis for the construction of individual implants such as cranioplasties often needed for covering cranium defects. They can also be generated with rapid prototyping processes. Such tailor-made implants are known as "taylored implants" (TI).

The data record of the 3D reconstruction is imported into a CAD system by reverse engineering processes. In the CAD system an implant is defined using design methods of mechanical engineering. This construction considers not only the basic geometric layout but also medical questions such as mounting, drainage, etc.

Prototypers and materials available at present are unable to produce directly implantable components. A positive has to be made in precision casting or alternatively in a mold that can be run in.

5.4.3   Characteristics of Medical Models

Medical models have a number of characteristics that are of secondary or of no importance for product development in mechanical engineering.

Large Amounts of Data

Data records of medical models contain large amounts of data owing to the density of raw data (e.g., the CT of a complete cranium in 1 mm layers contains approx. 150 layers à 512 · 512 pixels) and the complexity of the models. In comparison to mechanical engineering models they can contain up to 10 times the volume of data.

Large Models

Models for medical use often encompass more than the specific area of interest. For the medical treatment of all kinds of deformities, traumata, and large-area resections following a tumor disease especially, models are used not only for the limited local judgment of the defect and for introducing the optimal therapy including the production of possibly necessary implants, but they also need to fulfil aesthetic functions. With these models the physician can judge the results of his or her work on the general aesthetic. Therefore, it is usually necessary to have a model of the complete cranium, for example, and not just one of the immediate surroundings of the defect.

Several Models

If models are needed for planning-surgeries and trial-surgeries operations are carried out with them, a second model that documents the initial situation will usually be necessary. In future considerations concerning quality control will be taken into account.

Transparency

Stereolithography models are transparent, at least transluscent. Internal structures and hollow spaces (e.g., the mandibular nerve canal) are visible.

If transparency is required, the use of models made by other means is limited as they are not transparent and internal structures cannot be judged.

Ability to Be Sterilized

Planning surgery strategies and trial-surgeries, and the molding and fitting of implants makes sense only if the model can be taken into the operating theater. This can be done with stereolithography models of acrylates. Sinter models of nylon, polycarbonate, or polystyrene, however, are less suited as powder still contained in the internal hollows could escape when the model is used and because the models lose material constantly through abrasion as is the case with milling models of polyurethane foam. Such abrasion can be reduced by impregnating the model. 3D Printing models show similar behaviour. Therefore, these models cannot be sterilized in the strict sense. For the FDM process (see Section 3.3.4.1) there exists a material that can be sterilized.

Biocompatibility

Fully polymerized acrylate derived from stereolithography processes is considered the sole biocompatible material suitable for rapid prototyping. Epoxid resins are most certainly not biocompatible. As the material properties are at present insufficiently known it is not advisable to use an rapid prototyping models as direct implants. After all, toxic fumes are produced during the sintering of polystyrene. Sintered polystyrene, however, is not toxic.

Support Structures

The calculation and the optimal installation of support structures required in stereolithography and their expert removal, especially with complex structures such as the cranial base, demand certain medical knowledge of model makers.

Unconnected Model Parts

When bone structures are segmented, soft tissues such as connective tissues, cartilages, and muscles disappear from the data record, resulting in unconnected model parts. For medical therapy it is often very important to know the exact position of these unconnected model parts in relation to each other. With the aid of CAD programs, such parts are fitted with connecting clamps. In this way it is, for example, possible to display the exact position of the mandible joint in relation to the socket in the maxilla independently of the segmentation in the CT image.

Model definitions: For anatomic facsimile models are different from those in mechanical engineering definitions have to be adopted accordingly.

Geometrical Prototypes (Proportional Models)

These are used for visualizing 3D structures, especially bone structures, and facilitate a better diagnosis. They are also used for teaching and training medical personnel, especially with regard to pathological manifestations, for patient information, for documentation and, increasingly, also for quality control. They are used for developing and controlling surgery strategies and for preoperative planning.

Functional Prototypes (Functional Models)

These are the basis for trial-surgeries. Stereolithography material especially (acrylates) can be severed and drilled with tools usually available to physicians so that trial-surgeries can be carried out using common clinical devices. Stereolithography models have the additional advantage in their being able to be reassembled again with the aid of photosensitive resins and UV pistols, thereby enabling the same model to be used for various trial operations using a variety of strategies.

Anatomic facsimile models are also used for the preoperative fitting of implants.

The fitting of protheses is another area in which functional models are used. With the help of anatomic facsimile models it is also possible to follow the effects of long-term therapy, as is quite common, for example, in orthopedic cases, especially in orthodontics.

Lastly, functional models can be used for medical or biomedical research and they give, for example, geometrically exact reproductions of the present state of the patient that can then be integrated into larger model units (e.g., the reproduction of the larynx for airflow and vocal sound analyses).

Series Part (Implants)

Implantable models made on the basis of rapid prototyping strategies and cast implantable materials, for example, cranioplasty devices for covering larger defects, can be used directly.

In this way individual implants and similar biomedical products can also be produced. As such individual implants and protheses are needed only once, the functional prototype has the property of a "series part."

5.4.4   Example: Anatomic Facsimilies for a Reconstructive Osteotomy

Rapid prototyping task	Modeling a cranium for preoperative planning
Rapid prototyping process	Stereolithography
Model class	Geometrical prototype, functional prototype
Production Process	–

As a result of an inborn deformity a two-year old girl needed surgical rearrangement of the entire cranium geometry, known as craniosynostosis, was necessary. To prepare and plan the complicated surgery the deformity was at first reproduced as a 3D stereolithography model (Fig. 5-33). In the course of the preoperative planning the osteotomy lines were drawn in (Fig. 5-34). As the stereolithographic material can be worked on with tools that are usually available to physicians, the most important steps of a trial operation could be carried out. The condition after the osteotomy is illustrated in Fig. 5-35. By using osteosysthesis plates the cranial bones were newly configured (Fig. 5-36).

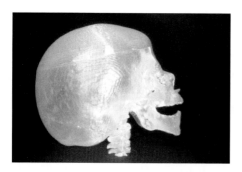

Figure 5-33 Craniosynostosis: original state, stereolithography model

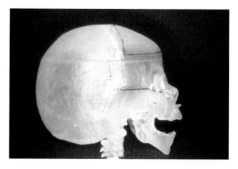

Figure 5-34 Craniosynostosis: osteotomy lines drawn in

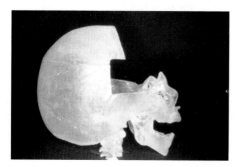

Figure 5-35 Craniosynostosis: condition after the osteotomy

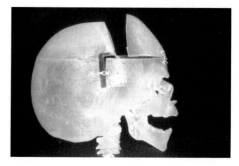

Figure 5-36 Craniosynostosis: newly confi- gured cranial bones. (Source: Bier, Charité, Campus Virchow- Klinikum, Berlin, Germany)

With the aid of such stereolithography models and their ability to make trial-surgeries possible, complicated surgical treatments can be planned effectively; the necessary informa- tion can be conveyed to the team, thereby enabling a faster and better quality surgery causing less stress for the patient.

5.4.5   Example: Covering of a Cranium Defect by Titanium Implant (Cranioplasty)

Rapid prototyping task	Production of a titanium implant for covering a cranium defect
Rapid prototyping process	Stereolithography selective laser Sintering
Model class	Series part
Production process (series)	Precision casting

A 22-year-old patient suffered a defect of 8 · 9 cm at the front of his skull caused by a severe traffic accident. This defect is too large to be covered by his own bone material. It was decided, therefore, to cover the defect by a suitable implant, a cranioplasty.

In the classical method, such implants were formed by hand, often on the open wound before being cast in implantable material. The difficulty lies in the intrinsic shrinkage factors of casting processes and their elimination by reworking the hand-shaped wax model. The extra stress the patient would suffer by a second surgery is also a factor to be borne in mind. Besides, hand-made implants are usually rather thick.

By applying the rapid prototyping processes stereolithography and laser sintering only one surgery was necessary and an implant was made immediately that fitted exactly.

In a first step, an actual size stereolithography model was generated on the basis of CT images. The cranium defect was reproduced with an accuracy of approx. ± 0.5 to 1 (Fig. 5-37).

In a further process step the STL points were regarded as elements of a point cloud and with the methods of reverse engineering were transformed as a continuous border of the defect. The border of the defect was then mirrored, thereby defining the outer edges of the implant for the defect. Using CAD programs, the spatially curved surface of the large outer periphery was constructed and an STL data record was generated. One rapid prototyping model made of nylon was produced on the scale 1:1 to be fitted into the cranium model; another one made of polycarbonate was produced that accounted for shrinkage occurring in the casting process. The polycarbonate model was molded into a titanium implant by means of a special precision casting process. The implant had such a perfect fit that it could be used without the brackets and drillings which are usually necessary. Drainage openings, occasionally used, were not installed in this case. Should these be necessary for medical reasons or if drillings for fixing devices are required, they are also easily incorporated into the model in the CAD.

The illustrations show the 3D reconstruction of the cranium defect (before removal of the artefacts) based on a spiral CT (Fig. 5-38), the stereolithography model with the hand-formed wax implant for planning the CAD construction (Fig. 5-39), the implant cast from the rapid prototyping positive, its extremely complicated 3D periphery (Fig. 5-40), and the titanium implant fitted into the stereolithography model (Fig. 5-41).

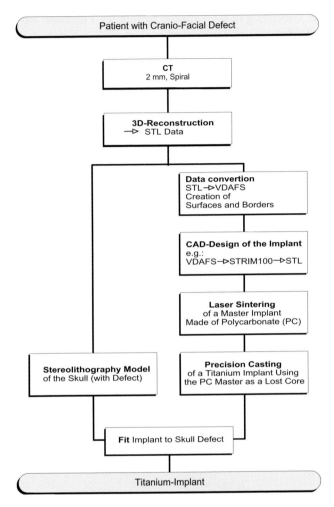

Figure 5-37 Process chain for the generation of a titanium implant

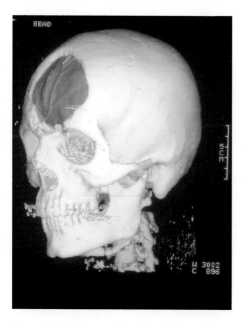

Figure 5-38 3D reconstruction of the cranium implant

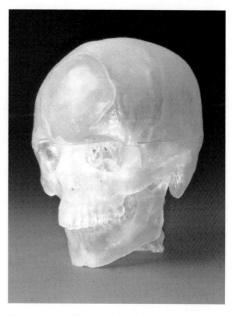

Figure 5-39 SL model with hand-formed wax

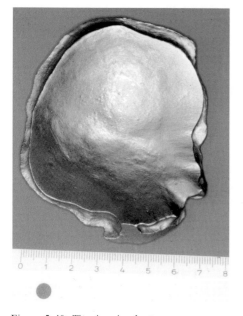

Figure 5-40 Titanium implant

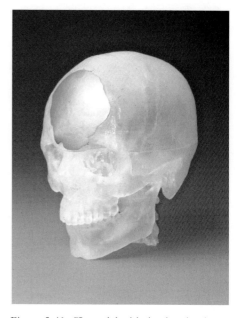

Figure 5-41 SL model with titanium implant. (Sources: Bier, Charité, Campus Virchow Klinikum, Berlin, Germany)

5.5 Rapid Prototyping in Art, Archaeology and Architecture

5.5.1 Model Making in Art and Design

For designers that are concentrating on styling the focus of interest lies on CAD systems that are able to support them in their creative work. It is important to keep in mind that designers want to use the computer for developing new shapes and generally playing around with it. Only if CAD systems are successfully developed in this direction and this aspect is sufficiently taken into account will stylists accept a computer as a replacement for their sketch book. The most important technical characteristic of such programs is the ability to work in absolutely real time. Another problem is to be found in the fact that stylists shape planes and bodies according to their imagination and not according to exact mathematical rules and principles. Rules and algorithms are required for translating the planes and volumes of the designer sketches into the (determined) CAD data familiar to industrial design engineers and that immediately report back the results to the stylist for checking.

Stylists concentrate on the appearance of a part. This occurs, for example, in respect to demands made on surface qualities and the effect of model surfaces. Designers who are used to appraising reflected lines on mirrorlike car bodies will hardly be able to accept the surfaces produced by current rapid prototyping processes.

5.5.2 Example Art: Computer Sculpture, Georg Glückman

Rapid prototyping task	Production of a physical model of a computer graphic
Rapid prototyping process	Laser sintering
Model class	Geometrical prototype
Production process	–

The artist Georg Glückman, who lives in Bochum and Vienna, started to work with CAD techniques in 1987. Since then he has been looking for methods and processes to enable him to reproduce his "computer sculptures" as physical 3D models. His first models were produced by 2½D milling from Styrodur and they were exhibited in the Blumenautomat-Galerie in Berlin, Germany, in the Pinx-Galerie, in Vienna, Austria, and on the Artware Fair in Hannover, Germany. In 1992, metal structures and tree structures were created that were produced by a computer-controlled punching machine. These artefacts were exhibited in the technical museum in Vienna, Austria, the Pinx-Galerie in Vienna, Austria, and on the Casa-Europea in Antwerp, Belgium.

During the same year (1992) the first stereolithography sculpture was created that was shown in exhibitions in Antwerp, Belgium, and Stuttgart, Germany. Figure 5-42 shows one of two 3D sculptures that are constantly penetrating one another and seem to float within one another. These sculptures were produced by a selective laser sintering process. They were presented in 1995 on the international Fair Triennale for small sculptures in Stuttgart, Germany, and Vienna, Austria.

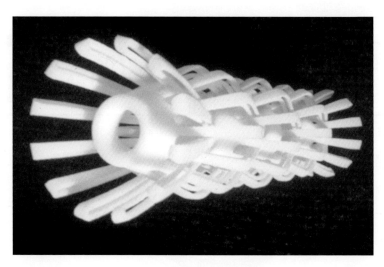

Figure 5-42   Computer-sculpture produced by selective laser-sintering, Georg Glückman 1995. (Source: LBBZ/CP)

5.5.3   Example Design: Bottle Opener

Rapid prototyping task	Production of a small series of bottle openers for the verification of the design and to support the decision for or against production
Rapid prototyping process	Stereolithography, vacuum casting
Model class	Technical prototype
Production Process	Laser cutting, injection molding

A bottle opener was to be designed as a gift for special customers. It should be easy to handle, have a pleasant and unassuming design, be attachable to a key ring, give the impression of precision, and be produced by a process employed in the rapid prototyping service company CP (Centrum für Prototypenbau GmbH, Erkelenz/Düsseldorf, Germany). The company colors and the logo had to be integrated into the design as discreetly as possible.

First designs were based on the letters C and P, the initials of the company. Figure 5-43 shows some of the first drafts. In the course of further development the design shown in Fig. 5-44 crystallized. It consists of a laser cut metal plate into which a cutout was incorporated in such a way that it could open bottles. In addition, the plate contains the company's initials and a loop for the key ring. A number of holes are drilled around the edge which on the one hand enable a secure and firm plastic coating to be made and on the other helps to avoid the plastic coating around the loop for the key ring from being ripped. The model was verified as a stereolithography model. As mentioned above the cutouts were made by laser.

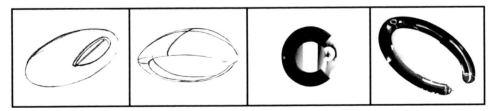

Figure 5-43 Bottle opener, first drafts. (Source: C. Schoenen/CP)

Figure 5-44 Bottle opener, final design. (Source: C. Schoenen/CP)

After several preseries had been produced by vacuum casting techniques, an aluminum mold was produced by high-speed milling and from this the required number of injection molding parts were produced. Figure 5-45 shows such a series part.

Figure 5-45 Bottle opener, series part. (Source: C. Schoenen/CP)

By using vacuum casting techniques, it was possible to test critical design elements such as the exact application of the relatively fine plastic rim, whether the hole for through which the key ring passes ripped easily and how far the device was bent when used, without having to produce a series injection mold. The decision for a design and thereby the decision for production was made solely on the basis of the vacuum cast prototype.

5.5.4   Example Archaeology: Bust of Queen Teje

Rapid prototyping task	Reproduction of the internal elements of the bust of Queen Teje which are not visible in the original
Rapid prototyping process	Stereolithography
Product Development Stage	Geometrical prototype
Production Process	–

The Egyptian Museum in Berlin, Germany, not only houses the world famous bust of Nofretete but also a rather unassuming bust of her mother-in-law, Queen Teje, 9.5 cm high. Her status changed from queen to goddess presumably when her husband, King Anophenes III, died. At the same time her royal headdress was covered by a hood made from papier maché set with blue pearls. After the x-ray technique had been discovered it was verified during the 1920s for the first time that the few golden elements that were visible through some slight damages in the outer hood really were part of an extensive gold headdress fitted onto a silver hood. Ever since it has been discussed whether the scientifically very interesting assessment of the headdress and the silver hood should be made possible by removing the also scientifically important outer hood which would be damaged at least by the procedure. When 3D modeling on the basis of a computer tomogram became better known it was decided to use this technology and to produce a 3D model of the bust beneath the hood with all its details.

A computer tomogram of the bust was taken in cooperation with the BAM (Bundesanstalt für Materialprüfung, Berlin) (the Federal Institute of Material Research, Berlin, Germany) and the data were processed in a stereolithography machine. In this case it was helpful that stereolithography enables the bust to be built in double size without difficulty so that some details could be better discerned. The data taken were separated mathematically so that the viscerocranium, the silver hood, and the jewelry could be built as single parts and the complete model could be built as an assembly of those single parts.

The results were received among experts with such enthusiasm especially with regard to their art historical and museum educational value, that the four stereolithography models are firmly established in their place right next to the original in the Egyptian museum in Berlin, Germany (Fig. 5-46).

Literature: [WIL95]

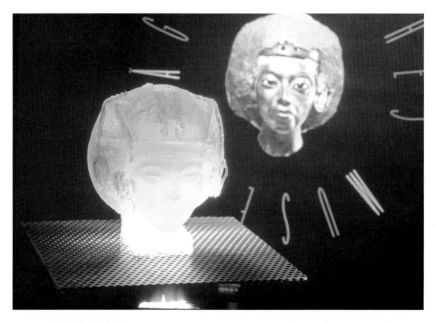

Figure 5-46 Stereolithography model of Queen Teje's bust with the bonnet removed, original.
(Source: Wildung, Egyptian Museum, Berlin, Germany)

5.5.5 Model Making in Architecture

In architecture and town planning models serve two main purposes. First, they help when solutions for special spatial-artistic designs are required and, here, the model construction is a kind of experimental medium for finding solutions. Second, at the end of a development process, they help to represent the architect's final design and give a clear picture of his or hers ideas. In both cases, the new methods of rapid prototyping are additional to the design repertoire. The ability to combine CAD with computer-supported model construction directly gives valuable support to the work of architects and town planners. The design process is not merely a matter of computer graphics in visual abstraction but also a produced design and perceptible object as known by architects and town planners from time immemorial.

The methods of rapid prototyping are used especially for complex design models in which internal geometrical and topological structures of buildings are important – for example, the layout of rooms inside a building – and that cannot be realized with other computer-aided model-making methods such as milling processes.

The use of rapid prototyping models and the necessary technical effort connected with it are worth the trouble only when complicated free-form areas or other architectural details must be reproduced true to life. Therefore, in architecture especially, it is preferred to use so-called "hybrid models" which are a combination of classical elements of architectural model construction and rapid prototyping elements.

5.5.6   Example Architecture: German Pavilion at Expo '92

Rapid prototyping task	Production of a complex, correct scale design model, highly detailed
Rapid prototyping process	Stereolithography
Model class	Concept model, geometrical prototype
Production Process	–

Figure 5-47 shows the design of the German Pavilion for the World Exhibition Expo '92 in Seville, Spain, which won first prize in the competition but which was not realized for cost reasons (design: Auer & Weber, Stuttgart/Munich, Germany).

The Expo pavilion on the scale 1:200 places high demands on the model-making process with regard to the geometry and the detail content owing to its extremely filigree support structures with their long slim supports and small profiled girders.

Figure 5-47   Design of the German Pavilion for the Expo '92 in Seville. Scale: 1:200. Stereolithography. (Source: B. Streich, University of Kaiserslautern, Germany)

5.5.7   Example Architecture: House Building Project

Rapid prototyping task   Correct scale reproduction of a house with filigree details
                                   and internal structures

Rapid prototyping process   Stereolithography

Model class   Concept model, geometrical prototype

Production process (series)   –

This is a good example of the application of computer-aided model-making processes in cooperative house building projects. It is based on an entry in a competition that took place during the South American architecture biennial in 1993 in Santiago de Chile (Fig. 5-48). The objective was, among other things, to develop types of buildings that would be especially beneficial to poorer people in densely built up areas.

The special challenge regarding the model making was to reproduce a model with all details exactly to correct scale (scale 1:100; model size: 90 · 90 ·70 mm). One of the vertical steel tubes in the main façade on the model, for example, has a diameter of 0.2 mm. It was also necessary to redesign the details within the hollow structures of the building exactly, such as dividing walls or the staircase at the far end of the building.

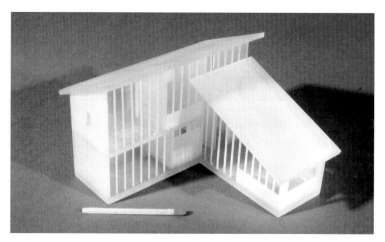

Figure 5-48   Design of a house for a housing construction project in Santiago de Chile. Scale: 1:100.
           Stereolithography. (Source: B. Streich, University of Kaiserslautern, Germany)

6 Economic Aspects

Providing clients with rapid prototyping models basically means selling time. If the client had time enough he would order it in some low-wage countries. Economic therefore equates to time saving.

Besides this, the question of economy in using rapid prototyping has two aspects: a strategic one and an operative one.

In considering the technology and its potential (Chapter 1), the strategic aspect is of considerable importance, that is, the question of whether rapid prototyping should be employed. The assessment depends to a large degree on the competitive strategy chosen and is not generally applicable.

From the operative standpoint it is to be decided which rapid prototyping process would be the best and in connection with which process chain it could be used to best advantage. The same considerations apply to service companies offering rapid prototyping and for rapid prototyping processes carried out in house.

Quite apart from all the other considerations, some technical criteria must be fulfilled simultaneously if rapid prototyping processes are to be applied economically:

- The models must be complex.
- The shortest possible development time is of utmost importance for the product's market position (in entertainment electronics this is a matter of months while the producers of heavy-duty engines need years).
- A volume-orientated 3D CAD system must be used as a matter of routine.

Even if only one of these criteria is not fulfilled, it must be assumed that the use of rapid prototyping is neither technically nor economically viable.

- The personnel available for the construction of rapid prototyping models must be committed and skilled.

6.1 Strategic Aspects

The fundamental decision to use rapid prototyping processes has economic consequences of varying importance, depending on the competitive strategies (see Chapter 1.1.3) chosen. When implementing the pioneer strategy, the cost leadership or the outpacing strategy, all critical success factors, are positively influenced by rapid prototyping. With the differentiating strategy, typical rapid prototyping factors such as time and money are of secondary importance. Therefore, the use of rapid prototyping shows no significant advantages if the differentiating strategy is employed.

First of all it has to be checked in all cases whether the use of rapid prototyping processes promises success. It is useful to go through a decision making matrix such as that shown schematically in Fig. 6-1. Even if the quantitative correlations required are not or not exactly known for specific cases the matrix assists in taking the important aspects of using rapid prototyping into consideration.

The starting point is the basic consideration (Chapter 1) according to which an economic optimization of the entire process must be achieved, preferably by minimizing product development time. Therefore, the key question must be (see Fig. 6-1):

Is it **possible** to shorten the product development time of my product significantly by using models?

The answer to this question depends basically on the product itself. A qualitative evaluation must be followed by a quantitative one. It needs to be established just how much time can be saved (at least) during each phase of product development by using models.

The second key question is:

To what extent will saving time by using models during product development really result in increased profitability of the product?

The answer depends basically on the market and especially on the product's life span. The effect is certainly higher for consumer goods than for capital goods. This also needs confirmation in figures.

These two quantitative relationships are difficult to establish, in specific cases it could even prove impossible. In these cases it would be better to make an estimate than to take over statements made in publications that are hardly verifiable (80% time saving is reported – often without mentioning what this evaluation is based on).

If models are already being used during product development, comparisons are available. The requirements the models have to fulfil are then already defined and a rapid prototyping process can be selected and evaluated as an alternative to the previously employed process.

If no model has been used previously, the requirements must first be established. In this case time comparisons do not exist. As model build times always need to be added to product development time, the time advantage can only be estimated by establishing the minimum time saving achievable. An increase in profit can be expected only if the difference between

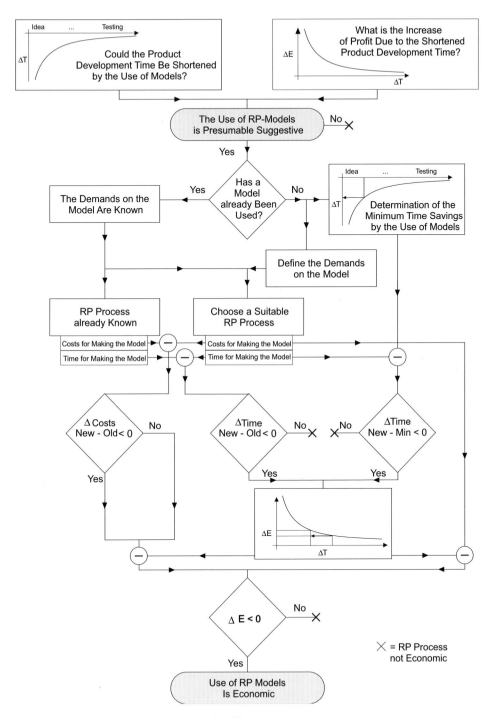

Figure 6-1 Economic analysis for the use of rapid prototyping

the actual build time and the time previously needed is at least the minimum time saving achievable.

Figure 6-1 shows the influence of time clearly. The question of whether the rapid proto-typing process is less expensive than the process previously used is merely a provisional estimate. Decisive is whether the gain in time can be realized as an increase in profit when the costs of making the model (or the increase in expenses for its construction) are also taken into account.

The observations show that the economic influence of rapid prototyping processes can be judged only if interactions with the entire product development process are included in the calculation.

For the quantitative evaluation of these effects reference can be made to Siegwart and Singer's model modified by Steger for rapid prototyping [STE95] (see Appendix A1 "Economy Model by Siegwart and Singer"). The critical success factors of time, quality, and flexibility are condensed into an efficiency value and from the factor expenses a cost value is established. The ratio of efficiency value to cost value gives a characteristic value for profitability WI and defines the economic potential of a specific technology. This value may be weighted according to probability, thereby further evaluating the risk to be taken if this technology be used (characteristic value of profitability WII).

The calculation is described in detail in the Appendix A1. Based on the assumptions made there the following characteristic values for profitability I and II result if the compatitive conventional process is "milling":

Strategy	WI / WI conventional	WII / WII conventional
Cost leadership	1.84	1.75
Differentiating strategy	0.86	0.70
Pioneer strategy	2.14	2.02
Outpacing strategy	1.91	1.76

With regard to cost leadership, pioneer strategy, and outpacing, the most important factors are time and expense – such strategies are effectively supported by rapid prototyping processes very nearly twice as effectively as by conventional processes.

The typical rapid prototyping advantages in time and money are of secondary importance with respect to the differentiating strategy. Quality and flexibility, however, factors that are not quite so positively supported by rapid prototyping, are of utmost importance. From this follows that the differentiating strategy is supported less by rapid prototyping than by conventional technologies

It is difficult to obtain exact figures since this calculation is also based solely on estimates, but nevertheless, dependences in the sense of a sensitivity analysis can be established by varying influencing factors which can then be taken into account.

6.2 Operative Aspects

Those working daily with rapid prototyping, external producers as well as internal rapid prototyping departments, must provide their services at market prices in competition with other rapid prototyping or conventional processes regardless of strategic considerations.

Therefore, they need criteria to assist in selecting the best process and in providing an economic evaluation.

The economic question has two main aspects:

- Which is the best rapid prototyping process to fulfil the requirements of quality and build time?

- What expenses are incurred by the use of rapid prototyping processes?

These questions already show clearly that there are no general answers; answers can be given only with regard to specific problem circumstances. For this reason, no attempt has been made to give a general solution. The factors supporting profitability are discussed instead.

6.2.1   Establishing the Optimal Rapid Prototyping Process or Processes

The fundamental influences concerning the selection of one specific process have already been discussed in previous chapters. The possibilities and limitations of single processes and their suitability for subsequent processes has been discussed in Chapter 2 in general and in Chapter 3 with regard to specialized machines. Technical data and material parameters are listed in the Appendices A3-1 to A3-14 and A3-15 to A3-25.

Accuracy	Chapters 2 and 3
Details	Chapters 2 and 3
Surfaces	Chapter 3 and Table 3-2,
Material	A2-15 to A2-25
Dimensions	A2-1 to A2-14
Mechanical-technological properties	A2-1 to A2-14
Subsequent processes, rapid tooling	Chapters 3 and 4

The necessary preparation and setup times such as preheating and inerting times for laser sintering, for example, account for a large part of the total time needed. Additionally there are time-consuming follow-up operations after the process is finished such as cooling phases (laser sintering), post-processings, and finishings, depending on the system used (e.g., varnishing of LLM models).

Profitability considerations made with regard to the speed of the rapid prototyping process itself must assume that the machines are running to capacity so that a basis is created for a fair comparison. It is one of the characteristics of rapid prototyping, however, that the models must be available within a short period of time and without, or with only a short, lead time. There is often little time left for planning machine utilization in the classical way. In addition, machine capacities must be kept free for such cases to ensure that the rapid prototyping processes itself does not become limiting factor in the product development chain.

The interrelation between the preparatory (CAD) work and the rapid prototyping process is especially important but is disregarded in all benchmark tests. The amount of work differs from model to model and depends also largely on the line of business, the CAD system, and the user. Therefore, the constant and optimal utilization of the prototyper and, consequently, its economic operation also depends largely on whether an optimal coordination exists between personnel and technical equipment of the preceding CAD, EDV, and data processing departments, and those of the subsequent rapid prototyping process.

The volume of the model is a decisive factor for the calculation of build time. It should be borne in mind that the build time increases with the height of the model (the volume remaining constant) since the processes are layer oriented and the recoating (i.e., the build progress in the z-direction) requires a certain period of time depending on the process. Therefore, the orientation of the models in the build chamber is of utmost importance.

In all processes requiring supports the period of time spent on post-processing is largely influenced by the type and number of the supports. In some cases it may be more economical to partly remove automatically generated supports manually before starting the build process than during the cleaning process. The final aim is a usable rapid prototyping model. Therefore, any calibrations and, if applicable, unsuccessful attempts must be included in the calculation.

Once the applicable rapid prototyping process or the processes have been defined the expenses may be calculated according to usual calculation practices (see [WIL95]).

6.2.2   Establishing the Costs of Rapid Prototyping Processes

The most important factors influencing costs are:

- Investment costs (Figs. 6-2 and 6-3)
- Additional investments
- Operating costs
- Expenses for service and maintenance
- Material expenses
- Operating expenses
- Personnel expenses

It is important to take into account not only the expenses for the machine itself but also the additional expenses such as the CAD and EDV accommodation, training, and installations for post-processing and finishing. These additional expenses including operating costs (hardly any prototyper can really be installed in an office – most need a workshop) add up to an average of 50% to 100% of the actual machine expenses.

The innovation cycles of prototypers are still very short. The expenses for updates (software) and upgrades (hardware) and for service and maintenance add up to at least 15% to 25% of the machine expenses.

A further factor affecting profitability is the question of whether each producer keeps his customer's machines up to date by means of upgrades, or whether the introduction of follow-on models necessitates the complete withdrawal from operation of the old models.

The large number of various formulations of maintenance contracts is especially confusing. Laser, scanner, updates, user group events, hot-line, and so forth are sometimes included, sometimes not, depending on the producer and the kind of contract. The expenses are difficult to compare in practice.

The material expenses of 50 to 130 $/kg (in the case of LLM processes, approx. 20 $/kg and Z-Corp, 20 $/kg) are significantly higher than those of materials used in cutting processes. However, no semifinished products are needed and no special tools or clamping elements are necessary. The materials, especially the expensive ones (photosensitive resins), may be used again but must be conditioned. In large sinter installations several thousand dollars are easily tied down by the powder material while the filling material for large stereolithography installations accounts for more than 50,000 $.

Operating costs include primarily power expenses for the laser, the heating and the expenses for gases (nitrogen, selective laser sintering) needed in the process, and for blasting material. Argon ion lasers especially are high-energy users and require additional installations for cooling due to their low efficiency (e.g., 11 kW input with 440 mW output). Newer devices such as solid-state lasers need much less energy. Processes not involving lasers and heating are even more economical.

The crucial work lies in the correct calculation of the time required and, consequently, the cost of the actual build process, especially initiation time, waiting time, and other ancilliary periods. Because this is done by taking measurements, it is mainly a question of diligence. The IWB-Anwenderzentrum Augsburg (IWB Application Center in Augsburg, Germany) established a detailed approach to calculation [WIL95].

The personnel expenses are largely similar to those of conventional model making processes. It should be remembered, however, that a CAD specialist is needed even when no 3D constructions are made in the prototyping sector. Furthermore, there is no clear qualification for an employee in the prototyping sector. Personnel must be trained internally or on self-organized training courses and fluctuation has to be accounted for.

The often vehement discussions concerning the speed of prototypers are leading in practice to not so controversial results, if not only the actual model construction in the process chamber of the machine is considered but also the entire model-making process starting with the data processing and including the post-processing. These are also the system limitations that are relevant to the user who wants to hold a model in his hands. All comparisons, therefore, that completely disregard the times needed for slicing and finishing, or standardize on a maximum or minimum time, must be rejected by the user.

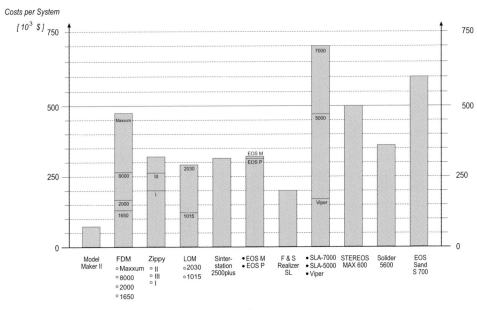

Figure 6-2 Standardized system expenses, functional prototypers

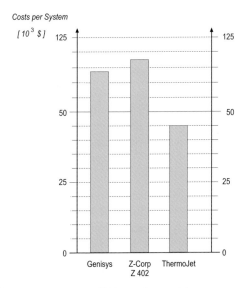

Figure 6-3 Standardized system expenses, solid imgaging prototypers

The two processes of stereolithography and selective laser sintering compete with one another especially in speed. Figure 6-4 shows a typical data set for mechanical engineering (relatively small) and a typical data set for medical use (relatively large) both built as models and analyzed. An epoxy resin was used in the stereolithographic process, while polyamide and polycarbonate were used for laser sintering. The examples show first that there is no great difference in total build times between stereolithography and laser sintering due to possible scatterings and second that the differences may be even greater within one machine concept because of the use of different materials (polyamide, polycarbonate), than between competing systems. The systematic shown in Fig. 6-4 is, however, well suited as a basis for the user's own analysis.

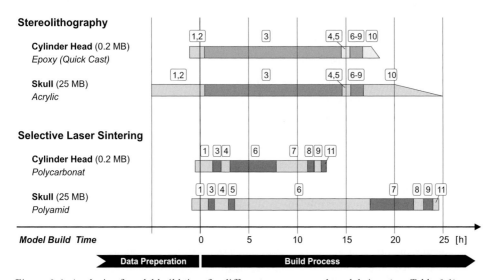

Figure 6-4   Analysis of model build time for different processes and model sizes (see Table 6-1). Source: LBBZ / CP

Table 6-1 Working steps corresponding to Fig. 6-4

Stereolithography	Selective Laser Sintering
1 Preheat laser	Fill in powder
2 Fill in resin	Put in base/bases if required
3 Build part	Flatten powder
4 Drain resin from part	Inert build space
5 Remove part from build space	Preheat build space
6 Clean part with TPM	Build part
7 Remove supports	Cool down build space
8 Finish part	Remove part from build space
9 Clean part	Remove loose powder
10 Post cure part	Remove supports/bases if required
	Finish part

6.3 Service

One aspect when considering profitability is the customer service situation. In connection with rapid prototyping this takes on a special meaning. Machines that are only "rapid" in theory do not help the user to shorten the "time to market" for his products. Unexpected downtimes that possibly overlap with troubleshooting are especially devastating in this area.

Customer service has been a point of contention over the last few years but has eased with regard to most producers today. Most local and foreign producers have established a viable service system and are still in the process of extending it successively. This should, however, not disguise the fact that travelling times are often clearly too long and that the engineer known as a specialist for that specific problem is usually not available.

A specific characteristic feature of rapid prototyping processes is that several sources of errors, for example, in stereolithography the laser, build-, and resin parameters, may be overlapping and appear as one big fault, so that only specific and time- and effort-consuming tests may isolate and eliminate the actual cause of the fault once and for all. Systematic troubleshooting – not a day-to-day routine but sometimes unavoidable – is made even more complicated as the suppliers of lasers and other components and the necessary know-how often is far away.

For these reasons it is advisable to check whether the most important replacements, including expensive elements such as lasers and scanners, should not be kept in stock. In the end this may prove to be the best alternative for the overall economy. The same is true for stock keeping of sufficient amounts of build material. The stocks should be quantified and reordered at fixed times so that extraordinary, inexplicable faults occurring after a change of material may be systematically traced by changing the material back to the one previously used.

6.4 Make or Buy?

The question of whether rapid prototyping should be applied at all (strategic aspect) as well as that regarding the application of the optimal process (operative aspect) depends largely on whether the models are made on the premises or whether they are bought from an external producer. The application of rapid prototyping processes on the premises has far reaching consequences concerning personnel and infrastructure quite apart from the question concerning the rate of utilization – an important aspect with all technologies. For these reasons, the pros and cons of making or buying are discussed in the following paragraph.

The following points speak for the use of rapid prototyping on the premises:

- If the machine is used to high capacity the whole profit can be realized.
- Competitive advantages can be realized by gaining specific know-how of rapid prototyping.

The following points speak against the use of rapid prototyping on the premises:

- Optimal technology can be applied in each case.

- External production is carried out regardless of capacity utilization and for previously established prices.

- A commitment to one or only a few processes is not necessary.

- Order peaks are easier to work off.

- Competition between external producers ensures that the latest technology is always available.

- Qualified personnel must be available and trained regardless of whether the machine is utilized or not.

- Extremely short innovation cycles force fresh investments before the machine is fully depreciated.

The absolute secrecy guarantee is essential for the external producer also and can be assumed to be as safe as if on the premises.

With so-called brokers, that is, agents for rapid prototyping services, who do not have their own installation, this presents more of a problem. Even if it can be assumed that they work with as much secrecy as the external producers with their own installation, their livelihood depends on the exploitation of temporal and structural price gaps. They can do that only if they spread their enquiries sufficiently, thereby losing direct control over the observance of secrecy; the risk of "leaks" increases.

7 Future Rapid Prototyping Processes

The speed with which rapid prototyping models have been developed from limited demonstration models to functional parts and preseries tool inserts has been breathtaking. The objective of making the transition from the prototype to the component, that is, from the prototype tool to the production tool, or, very generally, from the prototype process to the production process, has not yet been achieved. Since, from the rapid prototyping point of view, conventional processes are also being constantly developed further (see, e.g., the HSC process, Section 3.3.7.1) the objective recedes into the distance. Intensive research and development in industry, universities, and associations, however, have closed the gap slowly. This chapter shows with some examples which problems are being worked on and which solutions are favored.

The aim seems clear. Tool inserts and molds as well as metal functional parts are at the top of the list. Close consideration of rapid prototyping and its applications has, however, made apparent that the range of possible applications is far greater than was assumed and that its scope can probably not be assessed completely as yet. Further study is necessary. Research and development does not aim at one process nor can it be realized in one machine; rather it looks at systems of process-material-machine and their corresponding process chains.

Two mainstreams are recognizable: the development of new material systems and the qualification of new processes. These trends only point toward the main focus, but it is the entire process chain that needs consideration.

The selections in this book should be considered as examples and do not claim to be complete; above all, there is no intent to pass judgment.

7.1 Selected Trends in Material Development

Economically speaking rapid prototyping processes are especially effective if they enable the user to simulate the results of series processes and to optimize production accordingly in advance. Therefore, those materials that can be used in rapid prototyping processes and that allow a simulation of the relevant properties of the subsequent series product are especially important.

Furthermore, the material itself is often the object of development. In this case processes are needed that allow the simulation of material behavior. In view of the narrow range of rapid prototyping materials discussed in the preceding the known processes have their limitations.

This chapter concentrates on two selected examples of material development: the simulation of the production process for precision casting and the simulation of possible series materials.

7.1.1   Application: Precision Casting

The precision casting process is important for industrial scale manufacturing, which for profitable utilization requires a fully developed component that is optimized for easy casting. To this end models are needed that facilitate the fast production of series-identical precision casting parts. Current prototyping processes use rapid prototyping master models of plastic directly or they mold them in wax. The subsequent process is classical: repeated coating with a cream of ceramic, sanding, and finally the melting out of the master model and the baking of the mold.

The Fraunhofer-Institut für Produktionstechnologie (FhG-IPT), Aachen, (Fraunhofer Institute of Product Technology, Aachen, Germany) aims at the direct sintering of the ceramic mold in casting quality. To achieve this a special zircon mineral powder is sintered into thin-walled shells. On first sight this process is similar to the sand sintering process (see Fig. 3-39). Important is that there is no resintering. It therefore becomes a one-step process.

After loose particles have been removed from the finished model and it has been preheated it can be directly used for casting. Material developed for this process is of especially fine graining thereby enabling good reproduction qualities and accurcy (0.6% of the nominal size), and high-quality surfaces (RZ =30 to 50 µm) as required in precision casting.

The first examples (Fig. 7-1) produced in a trial installation show excellent results.

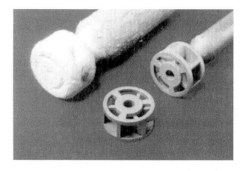

Figure 7-1 Directly sintered ceramic bowl, precision casting component. (Source: FhG-ILT)

The work is carried out on the basis of a commercial sinter machine (EOS). The further operations concentrate on industrially relevant scaling and the development of an industrial prototyper.

7.1.2 Application: Simulation of Series Materials

Rapid prototyping processes today are used mainly to simulate geometry and corresponding mechanical-technological properties, that is, the properties of the component. The subsequent series material is given. The properties of the prototypes should be as close as possible to those of the given material.

The materials themselves are becoming more and more the subject of optimization. It must be possible without having to make tools to test and optimize them with the aid of the prototype before series production starts. Usually, only very small amounts of material are available.

With the MJS process (see Section 3.3.4.2) a wide range of materials can be used, all of which stay viscous at temperatures of up to about 200 °C and can be extruded. Numerous possibilities are opened up with regard to the optimization of new materials:

- The simulation of materials for powder injection molding (PIM). With the MJS process it is possible to process directly powder-binder mixtures. The user can then study the behavior of debinder (e.g., process times, shrinkage) and sintering (e.g., shrinkage, density) and also the connected changes on the prototyper. The results thereby obtained on the component can be used for designing the mold and for optimizing the function.

- The simulation of materials for plastic injection molding. Apart from the ability of processing powder-binder mixtures, filled and unfilled plastics can also be processed.

- The simulation of filled waxes. For casting-technical processes, but also for medical use (e.g., anaplastology), the properties of differently filled waxes can be tested.

- The simulation of metals and ceramics. Powder-binder mixtures enable the production of prototypes from various metallic and ceramic material combinations.

- The process was tested on materials that are used for series today. After having been debindered and sintered, components made of stainless steel, for example (316L), achieve an tensile strenght of 480 MPa, while components made of ceramic (SiC) reach a flexural strenght of 310 MPa.

7.2 Selected Trends of Process Development

7.2.1   Laser Sintering of One-Component Materials

Metal models directly produced by rapid prototyping processes consist of material whose properties resemble, more or less, those of series materials which are, however, not used for series production. Their mechanical-technological properties, especially their dynamical parameters, are only partly known. Their porosity is very often far too high.

Because all processes dealing with the production of metal prototypes operate with the same layer orientated processes – those on which plastic prototyping processes are also based – one of the major problems of rapid prototyping processes remains: The surface qualities do not meet the expectations of the customers. Molds as well as models must undergo tedious rework operations before they can be used for their intended purpose.

Thinner layers only reduce the necessary finishing operations which again reduces the possibility of causing manual defects, but they do not principally solve the problem.

In spite of the great progress already made in the production of metal components, further research needs to be invested into the most important problems, primarily the improvement of surface qualities, the stability, the hardness, and, especially for tools, the overall life span.

Two process principles proved to be especially promising: the sintering in the powder bed and the generating process. The situation and the development trends are discussed in two examples.

7.2.2   Laser Sintering in a Powder Bed – Selective Laser Powder Remelting (SLPR)

7.2.2.1 Process Principle

Selective laser powder remelting (SLPR) is closely related to laser sintering in a powder bed. Laser sintering uses powders that are specially optimized for each process and therefore for a specific machine. This is also known as a powder system. Selective laser powder remelting processes can process common, that is, series-identical, powder materials, such as stainless steel, tool steel, or aluminum and titanium alloys.

The powder is processed directly without the addition of a binder. The components achieve a density of almost 100% already in the process. A subsequent infiltration process – usually used to increase the density and stabilty of porous components – is not necessary and is not possible owing to the high density.

To obtain components of such high density it is necessary to melt the powder particles completely. If this is not achieved and the particles are only incipiently melted, or connected with each other by sinter bridges, then the density of the components is not significantly higher than that of the powder bed. Because the process is carried out under atmosperic pressure, no significant mechanical precompression of the powder layer takes place.

Owing to surface tension there is a risk that the molten mass may form globular structures if the powder particles are completely melted. Because the laser beam melts several grains simultaneously these structures are usually larger than the powder grains. A hollow space is formed around the molten mass, as the density of the molten mass is higher than that of the surrounding powder. The single molten nodules are linked only at certain points, causing a porous structure in the component.

A special process control has been developed that prevents such globular structures being formed, especially in steel. By optimizing the process parameters and the protective gas control, and with the aid of a special scan strategy, it is ensured that the particles melted by the laser beam solidify in a dense, interconnected track (see Fig. 7-2).

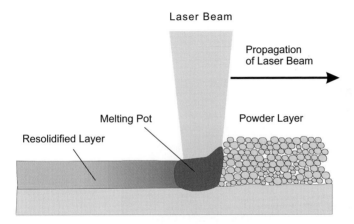

Figure 7-2 Schematic graph of the melting bath in selective laser Powder Remelting.
          (Source: FhG-ILT)

A closed layer is formed when adjacent tracks overlap. Because this area is remelted, adjacent tracks are fusion-metallurgically connected not only with one another but also with the subjacent layer.

7.2.2.2 Characteristics of Quality

Figure 7-3 shows the transverse microsection of a sample made of material 1.4404. The dark lines visible on the cross section mark the single working tracks within the melting bath parameters and demonstrate the overlapping of the adjacent and subjacent layers.

The transverse microsection also shows that the scan direction was turned by 90° after each layer. The high density and the fusion-metallurgic material interlockings ensure the high stability in the components produced. For establishing the stability the characteristic values of the tension test according to DIN EN 10002 are fixed for a proportional specimen with a thread head (Fig. 7-4) according to DIN 50 125 for the material example X2 CrNiMo 17 13 2. The tension test bars have a total length of 90 mm with a measurement length of 60 mm and a diameter of 8 mm.

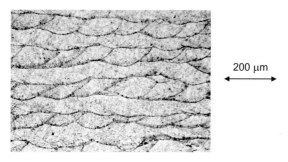

200 μm

Figure 7-3 Transverse microsection of a SLPR-sample made of stainless steel 1.4404.
(Source: FhG-ILT)

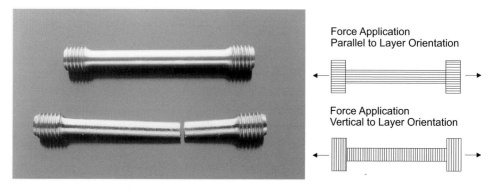

Force Application
Parallel to Layer Orientation

Force Application
Vertical to Layer Orientation

Figure 7-4 Geometry of the tension test bars and schematic graph of the power application relative to the layer generation. (Source: FhG-ILT)

To avoid notch effects caused by contractions at the layer joints the end contours of the tension test bars have been reworked by milling. To establish the influence the direction of the construction has on the stability, tension tests are carried out by applying the force parallel and vertical to the layers (see Fig. 7-5) [STA01].

The values established by tension tests of the yield strenght R_p 0.2%, the tensile strenght R_m and the elongation at break, are compared in Fig. 7-5 with the characteristic values for rolled sheet steel made of the same material. The yield strenght R_p 0.2 is significantly higher with 460 N/mm^2 or 404 N/mm^2 for sintered samples than the value of 190 N/mm^2 as given in the manufacturer's data sheets. However, the sintered samples show a value which is 56 N/mm^2 higher if the force is applied parallel to the tracks.

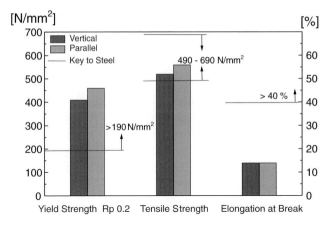

Figure 7-5 Results of tension tests for the material: 1.4404. (Source: FhG-ILT)

This behavior is also visible in the tensile strenght. It is assumed that defects between two layers are the reason for this phenomenon. If the force is applied vertically to the tracks, the specimen breaks at the defective joints; if the force is applied parallel to the layers, however, this joint is only an isolated fault within the stressed cross section plane.

The point of break confirms this assumption: While in all samples that were loaded parallel to the layers the breaking point lay in the middle of the tension length (point of greatest contraction), four out of five samples that were pulled vertically to the layers – built in the same build process – broke at the same point in the fringe area of the tension length. The overall tensile strenght lies within the range of 490 to 690 N/mm^2 as given in the manufacturer's data sheets. The elongation at break of the sintered samples lies at 14%, significantly below the value of 40% as given in the manufacturer's data sheets. The lower elongation at break is a result of hardening the material during the processing. The possible area of use for this process is determined not only by the density and stability but also by the achievable surface quality and the accuracy of the components. At present a surface roughness of $R_z < 60$ µm is achieved. The accuracy lies at about 0.1 mm.

The accuracy of the SLPR process is proven by the screw of the dimension M12 shown in Fig. 7-6. The screw was cleaned by sand blasting only to remove loose adherent particles

Figure 7-6 M12-screw produced by SLPR. (Source: FhG-ILT)

after completion of the construction process. Thereafter, it was possible without difficulty to screw a conventional nut onto the screw thread which was not processed further.

7.2.2.3 Examples of Applications

Owing to the quality characteristics of the components as described in the preceding and their ability to process various series-similar or series-identical materials, the SLPR is suitable for the production of metallic functional prototypes and directly useable components. Furthermore, the SLPR may be used for the production of tools that are used in injection molding and die casting either for prototypes or for small series.

Example: Pump impeller

The pump impeller shown in Fig. 7-7 were sintered directly of stainless steel in an SLPR process. The component shown on the right has protruding elements. These were realized with the aid of a support construction. Supports were also used to avoid warpings, for example. The supports were dimensioned in such a way that they could easily be removed after the build process.

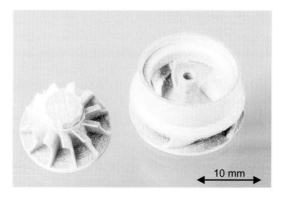

Figure 7-7 Functional prototypes produced by SLPR. (Source: FhG-ILT)

The build time for the functional prototypes shown in Fig. 7-7 was about 4h for the left-hand-side component and about 6h for the right-hand-side component. In addition it took another 0.5h to remove the supports on the right-hand-side component.

Example: Injection molding tool

Figure 4-10 shows a tool produced for plastic injection molding. The mold inserts were made by SLPR from material 1.4404. The construction took about 7h for producing the cavity and about 4h for the core.

Because in the SLPR process, the first layer is linked fusion-metallurgically to the substrate plate, this may be used as part of the model.

In this tooling method the substrate plate is used as the base plate for core and cavity, thereby shortening the total build time significantly. When comparing the actual build time for mold inserts of conventional production techniques (approx. 30h) with that of the SLPR (approx. 11h without reworking) it becomes clear that the significant potential of the SLPR process lies in decreasing the build times of complex injection molding tools.

Example:Die casting tool

The demands made on the tools used in a metal die casting process with regard to pressure and temperature are higher than for those used in a plastic injection molding process. Therefore, materials with higher resilience are used. A typical construction material for pressure die casting tools is the die steel 1.2343 (X38 CrMoV 5 1). To test the suitability of the SLPR for the production of pressure die casting tools a core slide was produced from this material (1.2343). The component reached a hardness of approx. 550 HV 0.3. The surface was reworked by polishing and the component was fitted into a series tool. After more than 1250 pieces were produced from the material GD-AlSi12 (230D) with a casting pressure of approx. 800 bar and a temperature of approx. 700 °C, the core was taken out to be examined. The core showed no wear due to stress.

Example: Light Metal

Highly stressed components are increasingly made of light metal such as aluminum and titanium. A sample component made of the aluminum alloy AlSi30 was produced by the SLPR process. The component showed a density of 100%.

The process is at present being optimized for other materials.

These examples show that the SLPR process enables the production of series-identical components in addition to the production of prototypes and prototype tools. With the ability to produce small series this process becomes a production process and thereby a process for rapid manufacturing.

7.2.3   Laser Generation

7.2.3.1 Principle of Process

Generation with the aid of laser radiation was derived from the single-layer laser coating which has become established as a production process in material processings. Figure 7-8 shows the principle of this process.

The material is a finely grained metal powder that is targeted onto a working area with the aid of an inert gas flow and a powder nozzle. There it is locally melted in the focus of the laser beam. The layer generation is achieved by solidification resulting from heat conduction into the component. Therefore, this is a one-step process.

Generation by laser radiation is a multilayer coating process (Fig. 7-9). 3D components are generated by stacking many layers on top of one another. During the process the additive materials and a thin boundary layer of the substrate are melted, thereby creating a metallurgic connection. The degree of mixture is adjustable by the use of parameters. Usually a finely grained texture results without cracks, bubbles, or pores with a density of 100% relative to the physical density of the starting material.

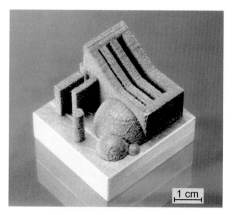

Figure 7-8 Single-layer coating with laser radiation

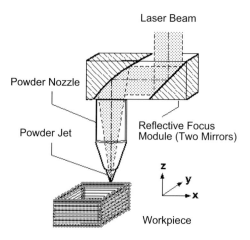

Figure 7-9 Generation by laser radiation

The most important process parameters are the laser power P_L, the feeding speed v_s, the powder mass flow m_P and the energy density distribution. Typical values are between 300 and 1000 W for P_L, between 300 and 1200 mm/min for v_s, and between 2 and 5 g/min for m_P. Since solids are generated as spirals or meanders in overlapping narrow tracks, the values specified for thin walls (wall thickness approx. 1 mm) are valid throughout.

A wide range of material is available for generation by laser radiation. Suitable materials are master alloys of iron, cobalt, copper, and nickel.

The energy density distribution and the method of powder feeding both influence the width and the height of the single track. The crucial factor is the specific energy density distribution, that is, the projection of the specific distribution within the beam onto the surface of the component. This varies with the movement of the beam over the substrate either in the x-direction or vertical to that direction because the beam cross section is elliptical owing to its positioning in translatory motion. CO_2, Nd:YAG and diode lasers are used.

This results, for example, in the formation of nonuniform layers along the perimeter of a generated cylinder. This effect can be avoided if the component is moved in a rotatory motion beneath a stationary laser beam and a stationary powder feeding nozzle.

It follows therefore that to achieve uniform results independently of the direction of the movement and the type of motion (translatory or rotatory) it is essential to have a homogeneous specific power distribution.

For most applications with single-layer coatings typical variations of the set value for layer thicknesses of 10 to 100m are tolerable. During generation these faults accumulate and can cause variations of several millimeters if 100 layers are generated.

Therefore, an exactly adjusted powder supply and a conveyance without temporal fluctuations in the powder gas beam are important requirements in addition to a homogeneous energy density distribution.

Generation requires higher qualities, therefore, than single-layer coating, especially with regard to the tools, laser beam, and powder gas jet.

Powder Nozzle Concepts

The material needed for the generation – in many cases a finely grained metal powder with grain sizes of 20 up to 200 μm – is transported by a powder conveyer through a feeding pipe to a powder nozzle.

Today pneumatic powder conveyors are usually used. The feeding gas takes over the transport and adopts the function of protective gas, that is, it protects the working area also against oxidation. Argon and helium are preferably used as protective gases. It is of utmost importance that the powder is evenly and continually conveyed and exactly proportioned, as even the slightest fluctuations in the flow of powder can already have negative results in the component.

The properties of the powder nozzle are, therefore, of special importance. For generation by laser radiation two different concepts of powder nozzles are principally suitable: the off-axis nozzle and the co-axis nozzle.

Off-axis nozzles are positioned outside the beam axis and are adjusted in such a way that the powder and the protective gas flow are available in the melting area.

The internal contour of the nozzle is usually cylindrical. When the gas flows out into the atmosphere a thin jet is formed which bursts after having travelled the distance of about five diameters and due to turbulences the effective amount of powder is reduced.

To achieve a high degree of powder efficiency a maximum working distance of approx. 10 to 15 mm must be maintained. The nozzles usually have an internal diameter of 2 to 3 mm. Such small distances cause a high heat load at the nozzle tip during generation. To avoid damage or occlusions the nozzles have an integrated water cooling system. The nozzles are made of copper because of its excellent heat conduction properties. The off-axis nozzle has the disadvantage that the quality of work depends on the direction in which the component is moved.

With co-axis nozzles, the powder is fed onto the working area co-axial to the laser beam. Therefore, the direction of movement does not influence the result. A differentiation is made between continuous and discrete co-axis nozzles.

Continuous co-axis nozzles distribute the powder evenly over the periphery and it is fed through a concentric gap around the laser beam. The relatively complicated internal construction of the co-axis nozzles poses a problem; internal powder blockages and subsequent periodic powder ejections can result.

In narrow areas that are difficult to reach (e.g., at the connecting rod bearing of a crankshaft) the relatively wide nozzle tip could cause problems during generation because it is extended laterally owing to the annular gap.

Continuous co-axis nozzles cannot be adjusted to differing focussing optics or to an energy density distribution that diverges from the rotation symmetry.

Discrete co-axis nozzles or multijet nozzles convey the powder through several single nozzles that are positioned around the laser beam. Often three single nozzles are positioned at 120° angles to one another. It is possible to focus the three single jets with a definition depth of several millimeters because the single nozzles are positioned at a relatively steep angle of 75° to the horizontal line and they can be adjusted independently.

Process Control and Adjustment

Depending on the geometry of the component local temperature peaks and subsequent construction faults may occur. This can be avoided by a temperature control. For this purpose a temperature range of 1400 °C–1700 °C is detected by means of a photodiode inside the melting bath and used as a signal for the laser power control.

Finally, a good well-defined process is achieved only when reproducible layer generation is ensured by the use of a distance control.

7.2.3.2 Quality Characteristics

The production of functional prototypes by generating requires that the model properties are as similar as possible to those of the subsequent series components.

The mechanical-technological properties of the generated components such as density, stability, hardness, and surface roughness are thererfore discussed in the following paragraphs.

Structure and Density

Figure 7-10 shows two micrographs of transverse sections of a thin-walled component under various degrees of enlargement. The very uniform layer construction is clearly recognizable. A finely grained texture with a dendritic structure is formed. No significant change is recognizable in the structure of the texture on the parting plane between two layers. The texture does not show any inclusions, pores, bubbles, or cracks. A quantitative density measurement by means of hydrostatic scales for a sample generated of stainless steel 316L showed a density of $\rho_{316L,\ generated} = 7.87$ g/cm^3 and is thus similar to the density of rolled sheet steel made of the same material (comparison value: $\rho_{316L,\ rolled} = 7.85$ g/cm^3). The density of the components in relation to the theoretical density of the starting material is therefore 100%.

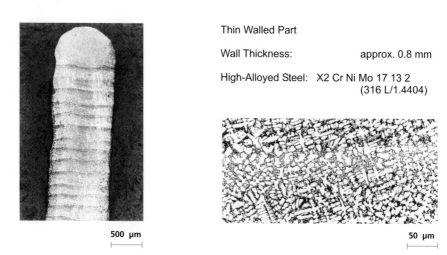

Thin Walled Part

Wall Thickness: approx. 0.8 mm

High-Alloyed Steel: X2 Cr Ni Mo 17 13 2
                                  (316 L/1.4404)

500 µm

50 µm

Figure 7-10 Micrograph of transverse section. (Source: FhG-ILT)

Hardness tests of thin-walled strands generated of stainless steel 1.4404 indicate an average value of 200 HV 0.3 and are, therefore, insignificantly higher than the specifications of the manufacturer's data sheets (120 to 180 HV). Hardness tests of rolled sheet steel made of the same material show an average hardness value of 160 HV 0.3. These measurements show that the samples generated by laser beam are on average slightly harder and slightly less ductile than rolled sheet steel made of the same material.

The static characteristic values taken from the tension test – yield strenght, tensile strenght, Young's modulus, and elongation at break – are shown in Table 7-1. To facilitate a comparison the characteristic values obtained from tension tests with rolled sheet steel and the data from the manufacturer's data sheets are entered for the corresponding characteristic values of stainless steel 1.4404.

Table 7-1 Mechanical properties of generated specimen of stainless steel (1.4404)

Property	Probe made by laser generation		Probe made by rolling		Key to steel [STA01]
	perpendic-ular	parallel	perpendic-ular to rolling direction	parallel to rolling direction	
Yield strength [N/mm²]	327 ± 53	367 ± 87	326 ± 17	342 + 30	> 190
Tensile strength [N/mm²]	495 +/– 139	437 +/– 25	639 ± 19	641 ± 22	490 – 690
Young's modulus [kN/mm²]	180 ± 23	164 ± 4	182 ± 30	160 ± 16	200
Elongation at break [%]	16 ± 13	7 ± 6	36 ± 3	42 ± 12	> 40

The values for the yield strenght $R_{P0,2}$ are very nearly identical for both generated samples and rolled sheet steel and are thus significantly higher than the 190 N/mm² given in the manufacturer's data sheets. The difference for generated samples is, depending on the direction of load, slightly higher, 40 N/mm², than that for rolled sheet steel, 16 N/mm². Owing to the higher standard deviation of the generated samples compared with rolled sheet steel this difference disappears in mean variation of test values. The yield strenght $R_{P0,2}$ is not significantly dependent on the direction.

More significant differences are to be found in the tensile strenght R_m. While the R_m for rolled sheet steel lies with approx. 640 N/mm² in the middle of the range as mentioned in the manufacturer's data sheets, the values of the generated samples lies with nearly 500 N/mm² (vertical to the layer generation) at the lower limit or, with approx. 440 N/mm² (longitudinal to the layer generation), outside the range. On average the tensile strenght of the generated samples in comparison to that of the sheet steel samples is about 75%. The tensile strenght of the generated samples does not depend on the direction; the average values of R_m differ by about 10%. In view of the high standard variations this is not significant.

The values for the Young's modulus are very nearly identical for both production processes and lie between approx. 160 kN/mm² and approx. 180 kN/mm², both being thereby slightly lower than those stated in the manufacturer's data sheets (200 kN/mm²). A significant dependence on the direction is not noticeable here either.

The greatest differences between the generated samples and the rolled sheet steel are found in the elongation at break. According to the manufacturer's data sheets the material should break only after a 40% elongation. Rolled sheet steel just achieves this value. Laser beam generated samples, however, break much earlier. If the differences in the elongations at break caused by high standard deviations are not taken into account, the values for generated samples are only one third of those for sheet steel samples.

Since the tensile strenght R_m also does not quite match the comparative values of sheet steel samples, while the yield strenght $R_{P0.2}$ and the Young's modulus E are very nearly identical with the comparative values of the sheet steel samples, it follows that the plastic ductility of laser beam generated components, as opposed to elastic tractility, is lower than that of rolled sheet steel. This points to an increase in hardness of the generated samples which is confirmed by hardness measurements.

The fatigue strength of generated samples (material 1.4404) is examined in bending tests (following DIN 50100; stress ratio R = 0.1, quasi-dynamic load, load frequency 30 Hz). The pulsating fatigue strength is established at 360 N/mm². The fatigue strength of the laser beam generated samples is therefore very similar to that of rolled sheet steel made of the same material.

Shape tolerances and positional tolerances of generated components are established according to DIN ISO 1101. With flatness, parallelity, and roundness, tolerances of 0.1 mm can be kept. The rectangularity tolerances are about 0.2 mm for solid volume components and about 0.4 mm for thin-walled components. These values are valid for CO_2 laser radiation (focussing optic: copper mirror, focus f = 250 mm).

Roughness measurements (waviness W_z and roughness R_z) were taken on strands of the material 1.4404. The waviness rectangular to the layers lies at nearly 30 to 70 μm, longitudinal to the layers between 13 and 35 μm, the roughness R_z between < 20 μm and 35 μm rectangular and about 10 μm longitudinal.

7.2.3.3 Examples of Application

Laser generation is used preferably when metal functional parts are needed that also possess complex geometries. Because the construction material and the energy necessary for the layer generation are provided solely by the movable laser nozzle unit, this process has full 3D capability and is not obliged to completely generate crosssection after cross-section of the component. In contrast to processes that first apply the material followed by the contouring of the layer, in this process the material can also be applied to arbitrary substructures or on to components that are to be repaired.

Example: Thin-walled Hollow Structures

Figure 7-11 shows thin-walled hollow structures generated by a laser beam having the geometric shape of turbine buckets. The single buckets have hollow walls and are generated from the material stainless steel 316L with a wall thickness of approx. 1 mm. A wall such as this they cannot be produced by milling processes. The buckets of the turbine wheel (on the right) were directly generated onto the hollow shaft. For this purpose the dimensions of the bottom layers had to be matched to the outer radius of the wheel (material: Ck 45).

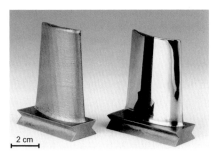

2 cm

Hollow Turbine Blade

Wall Thickness:
approx. 1 mm

Materials :   Steel
Shaft   :   Ck 45
Blades   :   1.4404

2 cm

Eight Turbine Blades Generated
Directly on the Shaft

Figure 7-11   Laser beam generated demonstration turbine buckets. (Source: FhG-ILT)

Both components demonstrate the potential of the laser generation, even if they are not functional parts of a turbine construction.

The total build time for the turbine wheel and single bucket was approx. 10h. The turbine is only partly suitable for fluidic studies owing to its surface roughness of $R_z > 50$ μm. The buckets would have to be polished for this purpose as shown for a single bucket in Fig. 7-11 (left).

Figure 7-12 shows the completely assembled demonstration turbine. The guiding grid consists of single buckets which are fixed in the turbine housing. Only the complex free form planes are generated by laser beam; all others were produced by cutting or were purchased in.

Example: Repair of a Turbine Bucket

Figure 7-13 shows the repair of a turbine bucket by means of laser beam generation while the process is in progress. The tip of the bucket wears away during use. The expensive buckets can be repaired several times, depending on the damage. A hollow-wall contour of about 3 to 5 mm in height and 1 mm ± 0.1 mm thick must be reconstructed.

The advantage of laser generation is that the construction is very similar to the final shape which enables the necessary finishings to be reduced to a minimum. Twenty-five single layers were generated, the total construction height is 5 mm.

Figure 7-12 Completely assembled demonstration turbine with generated wheel and generated guide blades. (Source: FhG-ILT)

Figure 7-13 Repair of a turbine bucket at the crown by laser beam generation. (Source: FhG-ILT)

Example: Car Gearbox Main Transmission Shaft

Figure 7-14 shows the wornout ball race on the main transmission drive shaft of a car gearbox, the repair work carried out by laser generation, and the repaired shaft before and after the finishing polish. To ensure the required resistance to wear, a cobalt basis alloy was chosen as layer material. Eight tracks with a lateral shift were generated in four layers one on top of the other.

Before Coating

After Coating

Car Axle Shaft

Material Deposition:
8 Tracks, Spiral
Side by Side in Each Layer
4 Layers One upon Each Other
Applied Material: Stellite 6

Process

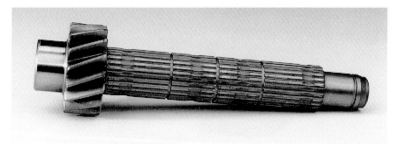

Final Shaft after Grinding

Figure 7-14 Repair of a car gear main transmission shaft by laser generation. (Source: FhG-ILT/ LBBZ)

Example: Injection Mold

A special field of application for the laser beam generation is the repair of injection molds. During use, and sometimes also during production, chippings and defects of all kinds occur that must be repaired speedily and as similarly to the original contour as possible. During the run-in of molds alterations are very often required which also demand the local application of material.

In such cases the generation offers the possibility of a well defined repair.

Figure 7-15 shows the mold insert of an injection mold for a reflector. The only difference of the further developed line of products to this older version is the sealing surface which is 4 mm lower. Accordingly, 4 mm of material is applied to the mold insert, thereby avoiding the production of a completely new mold. With a multijet nozzle the additional volume required is generated in 10 layers (layer thickness: approx. 0.5 mm) in the shape of a meander onto the old sealing surface. Subsequently, the mold is eroded to the final dimension and polished. The total time needed for this process is 5 h. The complete new production would take 3 days with three shifts per day.

Figure 7-15   Alteration of an injection mold with laser beam generation. Materials: Basic body 45 Ni Cr 6; construction Stellit 21. (Source: Fhg-ILT)

Appendix

A1 Economic Model by Siegwart and Singer

Siegwart and Singer developed a theoretical model for the calculation of the consequences of the application of new technologies. It leads to quantitative results taking into account the risk associated with the use of new technologies. Steger and Conrad [STE95] applied this model to rapid prototyping.

The following example shows how the model works in practice. It is based on the comparison of traditional five-axis milling and rapid prototyping by stereolithography.

Any evaluation of a new technique should take operational and strategic factors into account. Focusing first only on the operational point of view, the authors discover advantages for stereolithography in terms of speed and costs while the surface quality is better when using milling technologies. The strategic evaluation using Siegwart's and Singer's model does not only rely to the costs but takes all critical factors into account:

The Model

The practical application of the model leads to the calculation of a so called "economic efficiency coefficient" (EEC) that indicates to what degree rapid prototyping supports the investigated strategy. The economic efficiency coefficient is defined by the increase of performance (coefficient of performance, COP) divided by the costs (coefficient of costs, COC). The costs are obtained by adding all so-called traditional costs and the cost reduction due to rapid prototyping.

Performance is defined by the factors for success, which in our case are defined by costs, quality, and flexibility (in general, they might be defined in a different way according to the problem investigated). The calculation deals only with the differences (in % of the conventional case which is defined to be 100%).

Steger and Conrad applied this method of Siegwart and Singer to a strategic evaluation of the benefits caused by the use of rapid prototyping technology. The investigation was made for four different strategies of competition:

- Technology leadership (pioneer strategy),

- Cost leadership,

- Differentiation or concentration,

- Outpacing.

The influence rapid prototyping technology had on these strategies is entered into the model using the critical factors for success:

- Shortening of product development time,

- Reduction of costs,

- Increase in flexibility (product and production),

- Improvement of quality.

These critical factors for success are transformed into a coefficient of performance and a coefficient of costs, which both consist of two factors: the realization factor and the weight factor.

The realization factor is obtained by estimating the influence rapid prototyping will have on the critical factor of the traditional method used. If, for example, milling is the traditional procedure and the milling speed is the critical factor and therefore set to 100%, rapid prototyping will be estimated to speed up the fabrication process by 80%, leading to a realization factor of 180.

The weight factor indicates the importance of each criterion, it is set to 1, 2 or 3, (1 = less important; 2 = fairly important and 3 = very important).

The average value of the three coefficients of performance for time, quality and flexibility is then divided by the coefficient of costs. The result is the above mentioned economic efficiency coefficient, WI, and indicates the potential of a new technology.

There is a certain risk involved with every new technology, which is taken into account by multiplying each of the three coefficients of performance for time, quality and flexibility with a risk factor . The result is the economic efficiency coefficient WII.

Application to the introduction of Rapid Prototyping

Decreasing the product development time

The introduction of rapid prototyping is estimated to speed up production time by 80%. This can, depending on the strategy chosen, either be used for

- Earlier market entry,

- Increase of quality,

- Decrease of costs.

Whether there will be a definite decrease in the fabrication time or not, depends only on the strategy chosen.

Reduction of costs

The decrease of costs is based on the much cheaper fabrication of models and on the earlier possibility of judging the costs (see Section 1.3.2, Fig. 1-10). The savings due to the cheaper fabrication are estimated at 50% and those due to the earlier estimation of costs are estimated to another 25 %. Rapid prototyping mostly requires follow-up techniques which are not necessary when using traditional methods; therefore the estimated cost reduction may be lower.

The cost reduction can be realized in any case while the amount of realization regarding the other potentials depends merely on the chosen strategy.

Increase in flexibility (product and production)

When using rapid prototyping, the parts are available much earlier which enhances flexibility in general, on the market as well as within the company.

Changing market demands can be met by making parts identical to the later series in a short time. Shorter reaction time within the company improves communication in general, and enables the staff to fulfill changing requirements, *e.g.*, in the field of standardization or fabrication technology, quickly and reliably. This is also a precondition for livelong learning and therefore for a high innovation standard.

Improvement of quality

Rapid prototyping leads to a faster development of products and allows for the earlier evaluation of the product's properties (see Section 1.3.2, Figure 1-10). Methods for quality assurance and -control, such as quality function deployment (QFD), will therefore be supported effectively, thus leading to a better quality of the product. Later and very expensive corrections can be avoided.

This effect might not be decisive because quality assurance and -control methods are used in the field of traditional manufacturing as well.

Example

To reach comparable results, first a business strategy must be chosen, *e.g.*, "cost leadership". Second, a competing technology must be identified, *e.g.*, traditional milling.

For each critical factor for success, the realization factor using rapid prototyping has to be calculated, while each one for the traditional technology is set to 100% (= 100).

For example, a cost reduction of 75% due to rapid prototyping leads to a realization factor of 175. The strategy of cost leadership means that the time gained using rapid prototyping is completely "re-invested" to reduce costs. The other coefficients of success, development time and quality, therefore, remain unchanged (= 100). The increase of flexibility, which is difficult to estimate, was set at 120 (which means a 20% increase).

These realization factors are multiplied with the weight factors. For the cost leadership strategy, the weight factors are set to 3 for the costs, 2 for the time and 1 for each, quality and flexibility. Finally the average values for the coefficient of performance and the coefficient of costs are calculated, leading to the economic efficiency coefficient WI.

The risk associated with a new technology is put into the calculation by multiplying each coefficient with a risk factor (< 1). Then the average value of the coefficients of performance are evaluated and divided by the coefficient of costs. The result is the desired economic efficiency coefficient WII.

To achieve a quantitative result, first the values for the traditional technology are calculated. All risk factors are set to 1 or 100% accordingly; the following factors of economics, $WI_{traditional}$ and $WII_{traditional}$, result:

	$WI_{traditional}$	$WII_{traditional}$
Cost leadership	3.99	3.99
Differentiation or concentration	2.67	2.67
Technology leadership (pioneer strategy)	2.33	2.33
Outpacing	9.00	9.00

Next, the values for the application of rapid prototyping are calculated. The weight factors are calculated according to the strategy while the risk factors are independent from the strategy:

	Weight factor for			
	costs	time	quality	flexibility
Cost leadership	3	2	1	1
Differentiation or concentration	1	2	3	3
Technology leadership (pioneer strategy)	1	3	1	3
Outpacing	3	3	3	3
Risk factor	1.0	1.0	0.7	0.6

Dividing the values obtained for rapid prototyping technology by those for the traditional technology leads to the final results:

	WI/WI traditional	**WII/WII** traditional
Cost leadership	1.84	1.75
Differentiation or concentration	0.86	0.70
Technology leadership (pioneer strategy)	2.14	2.02
Outpacing	1.91	1.76

Values > 1 indicate that rapid prototyping technologies will successfully support the chosen strategy.

The more the success of a strategy is related to the factor time, the higher are the values obtained. The strategy of technology leadership or pioneer strategy is supported more then twice as effectively by rapid prototyping than by the traditional milling method. In contrast, differentiation or concentration strategies, which mainly rely on critical factors like quality or flexibility, are not better supported by the use of rapid prototyping technologies than by traditional milling techniques.

A2 Properties and Technical Data of Rapid Prototyping Systems and -Materials

This appendix summarizes all kinds of technical data and information on CAD-systems, Rapid Prototyping software, Lasers, Rapid Prototyping systems and materials for Prototypers and Cast Resin Tooling.

This information is based mainly on publications of manufacturers, on independent research institutes and on literature in general.

The data has been checked as good as possible, but due to frequent changes and re-labeling of products the user is requested to double-check before each application. To support this, the web pages are listed where ever possible.

In the Table A 2-4 to A 2-14 "Prototypers" there are as well some systems listed that are not discussed in detail in this book in order to perform some information on interesting but not widely spread prototypers.

Cast resin materials are available in uncountable variations and from various suppliers. The selection contains a flexible, a durable, a high temperature resistant and a crystal clear quality in order to show the wide range of applications. It does neither mean that the mentioned products are better than others available on the market nor does it represent some kind of a ranking.

Table	Content	Company
A2-1 A2-2	**CAD- and rapid prototyping software** 3D CAD systems Rapid prototyping software	
A2-3	**Laser for prototypers** Laser for prototypers	
A2-4 A2-5 A2-6 A2-7 A2-8 A3-9.1 A2-9.2 A2-10 A2-11 A2-12	**Prototypers** Manufacturers of prototypers Stereolithography systems I Stereolithography systems II Stereolithography systems III Sintering systems I Sintering systems II Sintering systems III Fused layer modeling systems I Fused layer modeling systems II Layer laminate manufacturing, LLM-Systems I	 3D Systems EOS F&S, Solider, Aaroflex 3D Systems (DTM) EOS F&S,Optomec FDM Solidscape Solidscape (LOM), Kira I
A2-13 A2-14.1 A2-14.2	**Prototypers** Layer laminate manufacturing, LLM-Systems II 3D printing systems I 3D printing systems II	 Kira II, Kinergy Z-Corporation, 3D Systems (Thermojet), Statasys (Dimension) Objet, Extrude Hone
A2-15 A2-16 A2-17 A2-18 A2-19 A2-20 A2-21 A2-22 A2-23 A2-24 A2-25	**Rapid prototyping materials and vacuum casting resins** Survey of photopolymers Physical properties of SL resins Mechanical properties of cured SL resins Thermal properties of cured SL resins Electrical properties of cured resins Properties of sintering materials I Properties of sintering materials II Properties of FDM materials Properties of ModelMaker materials Properties of LOM materials Properties of selected vacuum casting materials	

Table A2-1: 3D CAD systems

Manufacturer	Product	www
ALIAS/WAVEFRONT	Studiopaint 3D, Auto-studio Studio	www.aw.sgi.com
ANSYS Europe	Design Space	www.ansys.com
Autodesk Ltd	AutoCAD Mechanical Mechanical Desktop	www.autodesk.com
Bentley Systems	MicroStation	www.bentley.com
CADKEY Corporation	CADKEY Design Suite CADKEY Workshop	www.cadkey.com
Cimatron	CIMATRON E	www.cimatron.com
CoCreate Software	OneSpace Solution Suite OneSpace Designer	www.cocreate.com
Delcam	PowerShape ArtCAM	www.delcam.com
EDS PLM Solutions	I-DEAS Solid Edge	www.eds.com
EDS PLM Solutions	Unigraphics	www.eds.com
Evolution Computing	FastCAD	www.fastcad.com
Femsys	FEMGV 6.3	www.femsys.co.uk
Gibbs & Associates	GibbsCAM	www.gibbscam.com
Hewlett Packard	HP PE/Solid Designer	www.hp.com
IBM	CATIA V5	www.catia.ibm.com
ICEM CFD Engineering	ICEM CFD 4.2 ICEM DDN	www.icemcfd.com
IMSI	TurboCAD V8	www.turbocad.com
MASTERCAM	MASTERCAM	www.mastercam.com
Matra Datavision	EUKLID Strim 100	www.matra-datavision.com
MCS	ANVIL EXPRESS SNAP 2-3D	www.mcsaz.com
MSC.Software	MSC.Patran	www.mscsoftware.com
Pathtrace Engineering Systems	EdgeCAM	www.edgecam.com
PTC	CADDS 5i	www.ptc.com
PTC	Pro/ENGINEER Pro/DESIGNER	www.ptc.com
RealMation	Realimation	www.realimation.com
Roots Solutions	DESIGN WAVE	www.roots.co.uk
SolidWorks	SolidWorks 2001	www.solidworks.com
Tebis	Tebis	www.tebis.com
Toyota Caelum	Caelum II	www.caelum.co.jp
TranscenData Europe	CADfix	www.fegs.co.uk
Vero International Software	VISI-Modelling	www.vero-software.com
VX Corporation	VX V6.0	www.vx.com
X-NC	hyperFORM XCAM Desktop	www.xcad.co.uk

Table A2-2: RP-Software

Manufacturer	Actify Inc.	BIBA	DeskArtes Oy	Imageware Inc.	Materialise N.V.	Solid Concepts Inc.	ViewTec AG
www	actify.com	biba.uni-bremen.de	deskartes.fi	iware.com	materialise.com	solidcon-cepts.com	viewtec.ch
Software name	3D View	RP Workbench	Rapid Tools	RPM	Magics RP	Solid View/RP Master	Rapid View
RP-Software		yes	yes	yes	yes	yes	
CAD software / module	no	no		Surfacer (RPM optional)	no	no	no
Visualisation software	yes	yes			yes	yes	yes
Exchange format, Import	IGES, STL, VDA-FS, VRML, CATIA, ..	VDA-FS, STL, DXF, CLI, SLC	STL, VDA-FS, IGES	IGES, STL, DXF, VDA-FS, VRML, SLC	STL, DXF optional: VDA, IGES	IGES, VDA-FS, STL, VRML, 3D DXF,	STL, TDF, DXF, 3DS VRML
Exchange format, Export	VRML	STL, DXF, VRML, CLI, SLC, HPGL	STL, VDA-FS, IGES, CLI, SLI	IGES, STL, DXF, VDA-FS, VRML, SLC	STL, DXF, VRML, SLC, SSL, CLI, SLI	IGES, VDA-FS, STL, VRML, 3D DXF,	STL, TDF, DXF, 3DS VRML
Repair of STL data files	no	yes	yes	yes	yes	yes	yes
Manipulation of STL data files	no	yes	yes	yes	yes	yes	yes (optional)
Measurement of objects	yes	under Development	yes	yes	yes	yes	yes (optional)
Slicing of STL data files	no	yes	yes	yes	yes (optional)	yes (support only)	yes (optional)
Support generator	no	no	yes		yes (optional)	yes (optional)	yes (optional)
Tooling module	no	no	no	yes	yes (optional)	no	no
Operating system	Windows	Windows	Windows, Unix	Windows, Unix	Windows	Windows	Windows, Unix

Table A2-3: Laser for prototypers

	Dimension	Sintering	Stereolithography		
Laser		CO_2	He-Cd	Nd:YAG	Ar^+
Manufacturer		Synrad	Omnichrome	ADLAS / Coherent	328 Coherent
Type		45-5 Duo-Laser 2 · 25 W	3074-20M	421 QTII	Innova
Power	mW	60	20 (CW)	250 (10kHz Q switch)	440 (CW)
Wavelength	nm	1064	325	355	351
Mode		TEM_{00} statistically polarized	Multimode	TEM_{00}	TEM_{00}
Diameter of the unfocussed beam	mm	3.5	1.2	0.25–0.5	1.5
Divergence	mrad	5	1.2	2	1.5
Beam quality		$M^2 < 1.2$	$M^2 < 3.3$	$M^2 < 1.05$	$M^2 < 1.3$
Efficiency	%	~ 7	~ 0.5	0.2	~ 0.004

Table A2-4: Manufacturers of prototypers

Company	Headquarters	URL
3D Systems, Inc.	Valencia, California/USA	www.3dsystems.com
Aaroflex, Inc.	Manassas, Virginia/USA	www.aaroflex.com
Cubic Technologies, Inc.	Carson, California/USA	www.cubictechnologies.com
Cubital Ltd.	out of business since 2001	
DTM Corporation	taken over by 3D Systems in 2001	www.3dsystems.com
EOS GmbH Electro Optical Systems	Munich, Germany	www.eos-gmbh.de
Extrude Hone–ProMetal	Irwin, Pennsylvania/USA	www.extrudehone.com www.prometal-rt.com
F & S Stereolithographietechnik GmbH	Paderborn, Germany	www.fockeleundschwarze.de
Helisys, Inc.	out of business since 2001	
Kinergy Pte. Ltd.	Singapore	www.kinergy.com.sg
Kira Corp.	Aichi, Japan	www.kiracorp.co.jp
Objet Geometries, Inc.	Mountainside, New Jersey/USA	www.2objet.com
Optomec, Inc.	Albuquerque, New Mexico/USA	www.optomec.com
Sanders Prototype, Inc.	basic technology taken over by Solidscape	www.solid-scape.com
Schroff Development Corp.	Mission, Kansas/USA	www.schroff.com
Solidscape, Inc.	Merrimack, New Hampshire/USA	www.solid-scape.com
Soligen Technologies, Inc.	Northridge, California/USA	www.soligen.com
Stratasys, Inc.	Eden Prairie, Minnesota/USA	www.stratasys.com
Z Corporation	Burlington, Massachusetts/USA	www.zcorp.com

Table A2-5: Stereolithography Systems I

RP technology	Dimension	SL	SL	SL	SL
Prototyper name / type		SLA 3500	SLA 5000	SLA 7000	Viper si2
Manufacturer		3D Systems			
Equipment					
Dimensions	inch	37.5 / 40 / 78.3	74 / 47 / 79.5	74 / 64 / 80	52.5 / 33.5 / 70
Weight	lbs.	1350	2900	2630	1020
Power supply	V / A	200–230 / 15	200–230 / 15	200–230 / 20	115 / 15; 230 / 8
Power consumption (maximum)	kW	2.2	2.2	3	2
Operating gas	cu. feet/min				
Water supply	gal/min				
Extracting system capacity	cu. feet/min				
Chamber temperature	°F / °C	68–79 / 20–26	68–79 / 20–26	68–9 / 20–26	68–79 / 20–6
Relative humidity	%	< 50	< 50	10–50	20–50
Process					
Emissions (vapor, gas)		resin-dependent			
Disposal / household garbage		completely cured polymer			
Dangerous waste		liquid polymer, solvent			
Shaping element					
Technology		laser	laser	laser	laser
Type		Nd:YV04	Nd:YV04	Nd:YV04	Nd:YV04
x-y-contouring		scanner	scanner	scanner	scanner
Outline accuracy	inch	material-dependent	material-dependent	material-dependent	material-dependent
Repeat accuracy	inch				
Z-outline accuracy	inch	material-dependent	material-dependent	material-dependent	material-dependent
Z-outline repeat accuracy	inch	.0002	.0005	.0004	.0003
Recoating cycle time	s				
Recoating velocity	inch/s				
Power fluctuation	%	± 5	± 5	± 5	

Table A2-5 (cont.)

RP technology	Dimension	SL	SL	SL	SL
Part properties					
Maximum build volume	inch	13.8 x 13.8 x 15.7	20 x 20 x 23	20 x 20 x 23.62	10 x 10 x 10 / 5 x 5 x 10
Maximum part dimensions	inch	13.8 x 13.8 x 15.7	20 x 20 x 23	20 x 20 x 23.62	10 x 10 x 10 / 5 x 5 x 10
Layer thickness	inch	.002–.004	.002–.004	.001–.005	.002–.006
Beam spot size	inch	.008–.012	.008–.012	.009– .011/ .027– .033	.003 ± .0005/ .01 ± .001
Accuracy	inch/inch	.002	.002	.002	
Repeat accuracy	inch	.0002	.0005	.0004	.0003
Support structure		yes	yes	yes	yes
Interface					
Input data formats		STL, SLC	STL, SLC	STL, SLC	STL, SLC
Information processing system		PC	PC	PC	PC
Operating system		Windows NT	Windows NT	Windows NT	Windows NT
Software		Buildstation 5.0	Buildstation 5.0	Buildstation 5.0	Buildstation 5.2
Costs					
Prototyper (RP system)	US-$	350,000	500,000	799,000	179,000

Table A2-6: Stereolithography systems II

RP technology	Dimension	SL	SL	SL
Prototyper name / type		STEREOS 400	STEREOS MAX 600	STEREOS Desktop S
Manufacturer		EOS (Munich, Germany)–sales stopped in 1997		
Equipment				
Dimensions	inch	47.2 / 74.7 / 78.7	70.8 / 78.7 / 86.6	47.2 / 23.6 / 39.3
Weight	lbs.	1276	2860	484
Power supply	V / A	3 x 400 / 32	400 / 63	230 / 16
Power consumption (maximum)	kW	20		
Operating gas	cu. feet/min	compressed air (87 psi)		none
Water supply	gal/min	2.25–3	2.25–3	no
Extracting system capacity	cu. feet/min	3–11.8	3–11.8	1.8–11.8
Chamber temperature	°F / °C	64 – 86 / 18–30	64–86 / 18–30	63 – 81 / 17–27
Relative humidity	%	45	45	45
Process				
Emissions (vapor, gas)		resin-dependent		
Disposal / household garbage		completely cured polymer		
Dangerous waste		liquid polymer, solvent		
Shaping element				
Technology		laser	laser	laser
Type		Ar$^+$	(Nd:YAG), Ar$^+$	HeCd
x-y-contouring				
Outline accuracy	inch	± .002	± .002	± .002
Repeat accuracy	inch			
z-outline accuracy	inch	± .002	± .003	± .003
z-outline repeat accuracy	inch			
Recoating cycle time	s			
Recoating velocity	inch/s	7.9 (max.)	1.2–4.0	4.0 (max.)
Power fluctuation	%			± 3% / h
Part properties				
Maximum build volume	inch	15.7 x 15.7 x 9.8	23.6 x 23.6 x 15.7	9.8 x 9.8 x 9.8
Maximum part dimensions	inch	15.7 x 15.7 x 9.8	23.6 x 23.6 x 15.7	9.8 x 9.8 x 9.8
Layer thickness	inch	> .004	> .002	> .002
Beam spot size	inch	> .004	> .004	

Table A2-6 (cont.)

RP technology	Dimension	SL	SL	SL
Part properties				
Accuracy	inch/inch	.05–.1 %	.05–.1 %	.05–.1 %
Repeat accuracy	inch			
Support structure		yes	yes	yes
Interface				
Input data formats		STL, CLI, DXF, IGES, VDASF, Step		
Information processing system		PC		
Operating system		Windows		
Software				
Costs				
Prototyper (RP system)	US-$	no longer available		

Table A2-7: Stereolithography systems III

RP technology	Dimension	SL	SL (SGC)	SL
Prototyper name / type		FS-Realizer STL	Solider 5600	Solid Imager
Manufacturer		F&S Stereolitho-graphietechnik, Germany	Cubital Ltd., Israel out of business since 2001	Aaroflex
Equipment				
Dimensions	inch	31.5 / 31.5 / 90	161 / 67 / 59	8ft.10in./ 4ft.4in./ 7ft.4in.
Weight	lbs.	1100	13200	600
Power supply	V / A	230 / 16	3 x 400 / 40	120 / 20; 208 / 65
Power consumption (maximum)	kW	1.5	45	
Operating gas	cu. feet/min			
Water supply	gal/min			4
Extracting system capacity	cu. feet/min	yes		
Chamber tempera-ture	°F / °C	< 91 / < 33		68–80 / 20–27
Relative humidity	%	45		60–80
Process				
Emissions (vapor, gas)		low		
Disposal / household garbage		completely cured polymer liquid polymer, solvent		
Dangerous waste				
Shaping element				
Technology		laser		laser
Type		Nd-YAG		Ar^+, HeCd
x-y-contouring		scanner		Scanner
Outline accuracy	inch	± .002	.004	
Repeat accuracy	inch	.08 e-3		
z-outline accuracy	inch	± .004		.0025 (typically)
z-outline repeat accuracy	inch	± .004		
Recoating cycle time	s	< 10		30–75
Recoating velocity	inch/s	0–40		
Power fluctuation	%	±2		
Part properties				
Maximum build volume	inch	19.7 x 15.7 x 11.8	19.7 x 11.8 x 19.7	25 x 25 x 25
Maximum part dimensions	inch	19.3 x 15.4 x11.8	19.7 x 11.8 x 19.7	25 x 25 x 25

Table A2-7 (cont.)

RP technology	Dimension	SL	SL (SGC)	SL
Part properties				
Layer thickness	inch	> .004	.004–.01	.002–.020 (.005 typ.)
Beam spot size	inch	.004		.003–.008
Accuracy	inch/inch	± .004	.004	.0005
Repeat accuracy	inch	± .004		
Support structure		yes	–	yes
Interface				
Input data formats		STL, IGES, F&S	STL, IGES, CFL	STL, SLC, CLI
Information processing system		PC		Workstation
Operating system		Windows XP		Unix
Software		F&S		
Costs				
Prototyper (RP system)	US-$	200,000	no longer available	on demand

Table A2-8: Sintering systems I

RP technology	Dimension	SLS	SLS	SLS	SLS
Prototyper name / type		Sinterstation 2000	Sinterstation 2500	Sinterstation 2500plus	Vanguard HS si2
Manufacturer		3D Systems (DTM Corp.until 2001)			3D Systems
Equipment					
Dimensions	inch	114.2 / 59.1 / 74.8	114.2 / 59.1 / 78.7	106.3 / 53.1 / 78.7	83 x 51 x 75
Weight	lbs.	8800	8800	4743	3748
Power supply	V / A	3 x 400 / 40	3 x 400 / 40	240	3 x 240
Power consumption (maximum)	kW	6	7.5	7.5	12.5
Operating gas	cu. feet/min	N_2	N_2	N_2	
Water supply	gal/min	internal	internal	external	
Extracting system capacity	cu. feet/min	exhauster air			
Chamber temperature	°F / °C	84 / 29 (max.)	84 / 29 (max.)	84 / 29 (max.)	68 – 78 / 20 -26
Relative humidity	%	90 (non con.)	90 (non con.)	90 (non con.)	< 70
Process					
Emissions (vapor, gas)		yes (non-toxic)	yes (non-toxic)	yes (non-toxic)	
Disposal / household garbage		yes (non-toxic)	yes (non-toxic)	yes (non-toxic)	
Dangerous waste					
Shaping element					
Technology		laser	laser	laser	laser
Type		CO_2	CO_2	CO_2	CO_2
x-y-contouring		scanner	scanner + VBE	scanner + VBE	scanner
Outline accuracy	inch	± .0004 (scan. sys.)	± .0004 (scan. sys.)	± .0004 (scan. sys.)	
Repeat accuracy	inch	± .0004	± .0004	± .0004	
z-outline accuracy	inch	± .0004	± .0004	± .0004	
z-outline repeat accuracy	inch				
Recoating cycle time	s	8–10	8–10	5–7	
Recoating velocity	inch/s	variable	variable	variable	
Power fluctuation	%				

Table A2-8 (cont.)

RP technology	Dimension	SLS	SLS	SLS	SLS
Part properties					
Maximum build volume	inch	⌀ 12 x 15	15 x 13 x 16.5	15 x 13 x 18.1	14.5 x 12.5 x 17.5
Maximum part dimensions	inch	material-dependent	material-dependent	material-dependent	14.5 x 12.5 x 17.5
Layer thickness	inch	.002–.012	.002–.016	.002–.02	
Beam spot size	inch	.012–.02	.012–.02	.008–.016	
Accuracy	inch/inch	± .0051	± .0051	± .0051	
Repeat accuracy	inch	± .0047	± .0047	± .002	
Support structure					
Interface					
Input data formats		STL	STL	STL	STL, SLC
Information processing system		PC 586/90	PC 586/200	PC 586/330	PC
Operating system		Unix		Windows NT	Windows 2000
Software		DTM Application (Windows NT enables Magics RP)			
Costs					
Prototyper (RP system)	US-$	no longer available			250,000

Table A2-9.1: Sintering systems II

RP technology	Dimension	Direct metal laser sintering (DMLS)	Plastic sintering	Plastic sintering	Sand sintering (direct cast)
Prototyper name / type		EOSINT M 250 Xtended	EOSINT P 380	EOSINT P 700	EOSINT S 750
Manufacturer		EOS GmbH Electro Optical Systems, Germany			
Equipment					
Dimensions	inch	76.8 / 43.3 / 72.8	49.2 / 51.2 / 84.6	89.4 / 55.5 / 82.7	55.9 / 55.1 / 84.6
Weight	lbs.	2090	1760	4070	2816
Power supply	V / A	400 / 32	400 / 32	400 / 32	400 / 32
Power consumption (maximum)	kW	6 (average)	4 (average)	2.2 (average)	6 (average)
Operating gas	cu. feet/min	6 / comp. air (102psi)	11.8 / comp. air (87 psi)	11.8 / comp. air (87 psi)	8.83 / comp. air (87 psi)
Water supply	gal/min	0	0	0	0
Extracting system capacity	cu. feet/min	max. 58.8	11.8	118	23.5–88 (adjustable)
Chamber temperature	°F / °C	59–86 / 15–30	66–84 / 19–29	59–86 / 15–30	66–84 / 19–29
Relative humidity	%	max. 60 (at 68-77°F)	max. 50 (at 66-84°F)	max. 60 (at 68-77°F)	20–50 (at 66-84°F)
Process					
Emissions (vapor, gas)		low	low	low	low
Disposal / household garbage		sintered metal	powder residues	powder residues	cured sand
Dangerous waste		residual metal powder			sieving residue, phenolic resin
Shaping element					
Technology		laser	laser	laser	laser
Type		CO_2	CO_2	CO_2	CO_2
x-y-contouring		Galvoscanner	Galvoscanner	Galvoscanner	Galvoscanner
Outline accuracy	inch	.0002	.0004	.0004	.0004
Repeat accuracy	inch	± .0004	± .002	± .002	± .002
z-outline accuracy	inch	.0004	.0004	.0004	.0004
z-outline repeat accuracy	inch				
Recoating cycle time	s	variable	variable	variable	variable

Table A2-9.1 (cont.)

RP technology	Dimension	Direct metal laser sintering (DMLS)	Plastic sintering	Plastic sintering	Sand sintering (direct cast)
Shaping element					
Recoating velocity	inch/s	.4–4.0	2.4–4.0	4.7 (typ.)	4.7 (typ.)
Power fluctuation	%	process not affected by power fluctuation			
Part properties					
Maximum build volume	inch	9.8 x 9.8 x 7.9	13.8 x 13.8 x24.4	28.3 x 15.0 x 22.8	28.3 x 15.0 x 15.0
Maximum part dimensions	inch	9.8 x 9.8 x 7.3	13.4 x 13.4 x 23.6	27.6 x 14.6 x 21.7	28.3 x 15.0 x 15.0
Layer thickness	inch	.0008–.004	.004–.008	.004–.008	.008
Beam spot size	inch				
Accuracy	inch/inch	.002–.004 (typ.)	0.1–0.2 % (typ.)	0.1–0.2 % (typ.)	± .012
Repeat accuracy	inch				
Support structure		for overhangs > 45 °	no	no	no
Interface					
Input data formats		STL, CLI	STL, CLI	STL, CLI	STL, CLI
Information processing system		PC			
Operating system		Windows			
Software		EOS RP-Tools, EOS PSW, Magics RP/Materialise, 3Data Expert/Desk Artes			
Costs					
Prototyper (RP system)	US-$	335,000	315,000	735,000	690,000

Table A2-9.2: Sintering systems III

RP technology	Dimension	SLM–selective laser melting	Laser fused metal deposition	Laser fused metal deposition
Prototyper name / type		FS-Realizer SLM	Lens 750	Lens 850
Manufacturer		F&S Stereolitho-graphietechnik, Germany	Optomec	
Equipment				
Dimensions	inch	63.0 / 31.5 / 70.9	52 / 41 / 82	44 / 48 / 82
Weight	lbs.	1760	2500	3500
Power supply	V / A	380 / 16	3 x 208 / 75	3 x 480 / 75
Power consumption (maximum)	kW	3	20 (500W Laser)	40 (1000W Laser)
Operating gas	cu. feet/min	.106	Inert atmosphere–Ar, less than .5 cu.ft/min	
Water supply	gal/min	internal	6	13.2
Extracting system capacity	cu. feet/min		Ar recycling unit	Ar recycling unit
Chamber tempera-ture	°F / °C	68–77 / 20–25	50–105 / 10–40	50–105 / 10–40
Relative humidity	%	45	30–90	30–90
Process				
Emissions (vapor, gas)			purged Argon	
Disposal / household garbage			metal powder	metal powder
Dangerous waste				
Shaping element				
Technology		laser	laser	laser
Type		Nd-YAG	Nd-YAG	Nd-YAG
x-y-contouring		scanner	scanner	scanner
Outline accuracy	inch	± .002	.002	.002
Repeat accuracy	inch	.00008	.002	.002
z-outline accuracy	inch	layer thickness ± .002	.020	.020
z-outline repeat accuracy	inch	layer thickness ± .002	.020	.020
Recoating cycle time	s	10		
Recoating velocity	inch/s	0–40		
Power fluctuation	%	± 2	< 5	< 5
Part properties				
Maximum build volume	inch	9.8 x 9.8 x 9.8	12 x 12 x 12	18 x 18 x 42
Maximum part dimensions	inch	9.1 x 9.1 x 7.9	12 x 12 x 12	18 x 18 x 42

Table A2-9.2 (cont.)

RP technology	Dimension	SLM–selective laser melting	Laser fused metal deposition	Laser fused metal deposition
Part properties				
Layer thickness	inch	> .002	.010–.020	.010–.020
Beam spot size	inch	.006–.012	.040–.080	.040–.080
Accuracy	inch/inch	± .004	.001/1	.001/1
Repeat accuracy	inch	± .004	< .002	< .002
Support structure		yes		
Interface				
Input data formats		STL / IGES / F&S	STL	STL
Information processing system		PC	PC	PC
Operating system		Windows XP	Windows 2000/NT	Windows 2000/NT
Software		F&S	Proprietary Optomec Software	
Costs				
Prototyper (RP system)	US-$	350,000	(on demand)	750,000

Table A2-10: Fused layer modeling systems I

RP technology	Dimension	Fused deposition				
Prototyper name / type		Prodigy Plus	FDM2000	FDM3000	FDM Titan	FDM Maxum
Manufacturer		Stratasys				
Equipment						
Dimensions	inch	27 / 34 / 41	26 / 36 / 42	26 / 36 / 42	50 / 34.5 / 78	88 / 44 / 78
Weight	lbs.	300	396	396	1400	2500
Power supply	V / A	120 / 15; 230 / 7	120 / 20; 230 / 10	120 / 20; 230 / 10	3 x 230 / 16	230 / 32
Power consumption (maximum)	kW	1.5	1.5	1.8	2.8	5.2
Operating gas	cu. feet/min					
Water supply	gal/min					
Extracting system capacity	cu. feet/min					
Chamber temperature	°F / °C	65–76 / 18–24	65–76 / 18–24	65–76 / 18–24	65–84 / 18–29	65–84 / 18–29
Relative humidity	%	30–70	20–80	20–80	20–80	30–70
Process						
Emissions (vapor, gas)						
Disposal / household garbage		yes	yes	yes	yes	yes
Dangerous waste						
Shaping element						
Technology		extruder	extruder	extruder	extruder	extruder
Type						
x-y-contouring		plotter	plotter	plotter	plotter	MagnaDrive
Outline accuracy	inch	.001	.001	.001	.001	.001
Repeat accuracy	inch					
z-outline accuracy	inch	.004	.004	.004	.004	.004
z-outline repeat accuracy	inch	.0006	.0006	.0006	.0006	.0006
Recoating cycle time	s	0	0	0	0	0

Table A2-10 (cont.)

RP technology	Dimension	Fused deposition				
Shaping element						
Recoating velocity	inch/s					
Power fluctuation	%	0	0	0	0	0
Part properties						
Maximum build volume	inch	8 x 8 x 12	10 x 10 x 10	10 x 10 x 16	14 x 16 x 16	23.6 x 19.7 x 23.6
Maximum part dimensions	inch	8 x 8 x 12	10 x 10 x 10	10 x 10 x 16	14 x 16 x 16	23.6 x 19.7 x 23.6
Layer thickness	inch	.007–.013	.004–.020	.002–.030	.007–.010	.007–.010
Beam spot size	inch		.01–.1	.01–.1		.015–.020
Accuracy	inch/inch	± .008	± .004	± .005	± .005 / ± .0015	± .005 / ± .0015
Repeat accuracy	inch	.001	.001	.001	.001	.001
Support structure		yes	yes	yes	yes	yes
Interface						
Input data formats		STL	STL	STL	STL	STL
Information processing system		PC, Work-station	PC, Work-station	PC, Work-station	PC, Work-station	PC, Work-station
Operating system		Windows 2000/XP/NT	Windows 2000/XP/NT	Windows 2000/XP/NT	Windows 2000/XP/NT	Windows 2000/XP/NT
Software		Insight	Insight	Insight	Insight	Insight
Costs						
Prototyper (RP system)	US-$	63,000	no longer available	120,000	190,000	$ 250,000

Table A2-11: Fused layer modeling II

RP technology	Dimension	Fused layer modeling	
Prototyper name / type		ModelMaker II	PatternMaster
Manufacturer		Solidscape	
Equipment			
Dimensions	inch	34 / 26 / 59	34 / 26 / 59
Weight	lbs.	340	340
Power supply	V / A	115 / 20; 230 / 15	115 / 20; 230 / 15
Power consumption (maximum)	kW		
Operating gas	cu. feet/min		
Water supply	gal/min		
Extracting system capacity	cu. feet/min		
Chamber temperature	°F / °C	60–80 / 16–27	60–80 / 16–27
Relative humidity	%		
Process			
Emissions (vapor, gas)			
Disposal / household garbage		waxes, thermoplastic	
Dangerous waste			
Shaping element			
Technology		inkjet print head	inkjet print head
Type			
x-y-contouring		plotter	plotter
Outline accuracy	inch	± .001	± .001
Repeat accuracy	inch		
z-outline accuracy	inch	± .001	± .001
z-outline repeat accuracy	inch		
Recoating cycle time	s		
Recoating velocity	inch/s		
Power fluctuation	%		
Part properties			
Maximum build volume	inch	12 x 6 x 8.5	12 x 6 x 8.5
Maximum part dimensions	inch	12 x 6 x 8.5	12 x 6 x 8.5
Layer thickness	inch	.0005–.003	.0005–.003
Beam spot size	inch	.003	.003
Accuracy	inch/inch	± .001	± .001
Repeat accuracy	inch		
Support structure		yes (wax)	yes (wax)
Interface			
Input data formats		STL, SLC, DXF	STL, SLC, DXF
Information processing system		PC	PC

Table A2-11 (cont.)

RP technology	Dimension	Fused layer modeling	
Interface			
Operating system		Windows 95/98/NT	Windows 95/98/NT
Software		ModelWorks	ModelWorks
Costs			
Prototyper (RP system)	US-$	66,900	76,900

Table A2-12: Layer laminate manufacturing, LLM-Systems I

RP technology	Dimension	LOM	LOM	PLT–paper lamination technology	
Prototyper name / type		LOM-1015Plus	LOM-2030H	PLT-A4	PLT-A3
Manufacturer		Cubic Technologies – formerly known as Helisys		KIRA Corporation	
Equipment					
Dimensions	inch	48.4 / 29.1 / 51.6	55.5 / 81 / 55	33.3 / 34.1 / 47.4	45.1 / 31.4 / 48.2
Weight	lbs.	1000	2841	990	1210
Power supply	V / A	2 x 110 / 20	230 / 30	100 / 15	3 x 200 / 30
Power consumption (maximum)	kW				
Operating gas	cu. feet/min				
Water supply	gal/min				
Extracting system capacity	cu. feet/min	250	500		
Chamber temperature	°F / °C	68–80 / 20–27	68–80 / 20–27	50–86 / 10–30	50–86 / 10–30
Relative humidity	%	< 50		35–60	35–60
Process					
Emissions (vapor, gas)		yes	yes		
Disposal / household garbage		yes	yes	yes	yes
Dangerous waste					
Shaping element					
Technology		laser	laser	cutter	cutter
Type		CO_2	CO_2		
x-y-contouring		plotter	plotter	plotter	plotter
Outline accuracy	inch	.001	.001	± .002	± .002
Repeat accuracy	inch	.002	.002		
z-outline accuracy	inch			± .004	± .004
z-outline repeat accuracy	inch	layer thickness	layer thickness		
Recoating cycle time	s				
Recoating velocity	inch/s				
Power fluctuation	%				
Part properties					
Maximum build volume	inch	10 x 15 x 14	22 x 32 x 20	11 x 7.5 x 7.9	15.7 x 11 x 11.8
Maximum part dimensions	inch	10 x 15 x 14	22 x 32 x 20	11 x 7.4 x 7.8	15.7 x 11 x 11.8

Table A2-12 (cont.)

RP technology	Dimension	LOM	LOM	PLT–paper lamination technology	
Part properties					
Layer thickness	inch	.003–.008	.003–.008	.003 / .006	.003 / .006
Beam spot size	inch	.008–.010	.008–.010		
Accuracy	inch/inch	± .006	± .006	± .008	± .008
Repeat accuracy	inch				
Support structure					
Interface					
Input data formats		STL	STL	STL / RPF / RPS / JAMA-IGES	
Information processing system		PC	PC	PC	PC
Operating system		Windows NT	Windows NT	Windows 95/NT	Windows 95/NT
Software		LOMSlice	LOMSlice	RP-CAD	RP-CAD
Costs					
Prototyper (RP system)	US-$	69,950	175,000	62,000	95,000

Table A2-13: Layer laminate manufacturing, LLM-Systems II

RP technology	Dimension		RPS	RPS	RPS
Prototyper name / type		PLT-A3	ZIPPY I	ZIPPY II	ZIPPY III
Manufacturer		Kira Corporation	Kinergy PTE Ltd, Singapur		
Equipment					
Dimensions	inch	45.1 / 31.4 / 48.2	68.1 / 39.4 / 63	101.2 / 73.2 / 78.7	82.7 / 59.1 / 70.9
Weight	lbs.	1210	1760	3300	5000
Power supply	V / A	3 x 200 / 30	220 / 20	380 / 30	380 / 25
Power consumption (maximum)	kW		4.4	11.4	9.5
Operating gas	cu. feet/min				
Water supply	gal/min		2	2	2
Extracting system capacity	cu. feet/min		5.3	7.1	7.1
Chamber temperature	°F / °C	50–86 / 10–30	68–82 / 20–28	68–82 / 20–28	68–82 / 20–28
Relative humidity	%	35–60	60	60	60
Process					
Emissions (vapor, gas)					
Disposal / household garbage		yes			
Dangerous waste					
Shaping element					
Technology		cutter	laser	laser	laser
Type			CO_2	CO_2	CO_2
x-y-contouring		plotter	Cartesian Robot	Cartesian Robot	Cartesian Robot
Outline accuracy	inch	± .002	.0004	.0004	.0004
Repeat accuracy	inch		.0004	.0004	.0004
z-outline accuracy	inch	± .004	.0004	.0004	.0004
z-outline repeat accuracy	inch		.0004	.0004	.0004
Recoating cycle time	s				
Recoating velocity	inch/s				
Power fluctuation	%				

Table A2-13 (cont.)

RP technology	Dimension		RPS	RPS	RPS
Part properties					
Maximum build volume	inch	15.7 x 11 x 11.8			
Maximum part dimensions	inch	15.7 x 11 x 11.8	15.7 x 11.8 x 13.8	46.5 x 28.7 x 21.6	
Layer thickness	inch	.003 / .006	.005	.005	.005
Beam spot size	inch		13.8	34.6	23.6
Accuracy	inch/inch	± .008	.006	.012	.01
Repeat accuracy	inch		.006	.012	.01
Support structure					
Interface					
Input data formats			STL	STL	STL
Information processing system		PC	PC	PC	PC
Operating system		Windows 95/ NT	Windows 98	Windows 98	Windows 98
Software		RP-CAD	RPP-S016	RPP-S026	RPP-S036
Costs					
Prototyper (RP system)	US-$	95,000	price on demand		

Table A2-14.1: 3D printing systems I

RP technology	Dimension	3D printing					3D printing	3D printing
Prototyper name / type		Z 402	Z 402 C	Z 400	Z 406	Z 810	ThermoJet	Dimension
Manufacturer		Z Corporation					3D Systems	Stratasys
Equipment								
Dimensions	inch	40 / 31 / 44	40 / 31 / 44	29 / 36 / 42	40 / 31 / 44	95 / 45 / 76	54 / 30 / 44	27 / 36 / 41
Weight	lbs.	330	330	300	470	1240	826	300
Power supply	V / A	230 / 16	230 / 16	230 / 16	100–230 / 4	115 / 20; 230 / 12	100/12.5; 115/10; 230/6.3	115 / 15
Power consumption (maximum)	kW	< 1	< 1	< 1	< 1	< 1		
Operating gas	cu. feet/min							
Water supply	gal/min							
Extracting system capacity	cu. feet/min							
Chamber temperature	°F / °C	65–77 / 18–25	65–77 / 18–25	65–77 / 18–25	65–77 / 18–25	65–77		65–77 / 18–25
Relative humidity	%	30–80	30–80	30–80	30–80	30–80		30–70
Process								
Emissions (vapor, gas)								
Disposal / household garbage		yes	yes	yes	yes	yes		
Dangerous waste								

Table A2-14.1 (cont.)

RP technology / Shaping element	Dimension	3D printing					3D printing	3D printing
Technology		1 inkjet print head	1 inkjet print head	1 inkjet print head	4 inkjet print heads	6 inkjet print heads	print head	print head
		Canon BubbleJet			HP–DeskJet			
Type		printer / plotter	printer / plotter	printer / plotter	printer / plotter	printer / plotter	printer / plotter	printer / plotter
x-y-contouring								
Outline accuracy	inch	180dpi	180dpi	180dpi	300dpi	300dpi	300 / 400 dpi (x-y)	
Repeat accuracy	inch	180dpi	180dpi	180dpi	300dpi	300dpi		
z-outline accuracy	inch						600dpi	
z-outline repeat accuracy	inch							
Recoating cycle time	s	3	3	3	3	5		
Recoating velocity	inch/s							
Power fluctuation	%							
Part properties								
Maximum build volume	inch	8 x 10 x 8	8 x 10 x 8	8 x 10 x 8	8 x 10 x 8	20 x 24 x 16	10 x 7.5 x 8	8 x 8 x 12
Maximum part dimensions	inch	8 x 10 x 8 (monochrome)	8x10x8; color mode: 6x9x8	8 x 10 x 8 (monochrome)	8 x 10 x 8 (mono-chrome+color)	20x24x16 (mono-chrome+color)	10 x 7.5 x 8	8 x 8 x 12
Layer thickness	inch	.003–.010	.003–.010	.003–.010	.003–.010	.003–.010		.010 / .013
Beam spot size	inch	.016 (monochrome)	.016 (mono-chrome)	.016	.087 (mono-chrome)	.13 (mono-chrome)		

Table A2-14.1 (cont.)

RP technology	Dimension	3D printing	3D printing	3D printing	3D printing	3D printing	3D printing	3D printing
Part properties								
Accuracy	inch/inch	0.4 %	0.4 %	0.4 %	0.4 %	0.4 %		± .01
Repeat accuracy	inch							
Support structure								(automatically)
Interface								
Input data formats		STL	STL	STL	STL / VRML / PLY / SFX	STL / VRML / PLY / SFX		STL
Information processing system		PC	PC	PC	PC	PC	PC	Workstation
Operating system		Windows 2000/NT	Windows 2000/NT	Windows 2000/NT	Windows 2000/NT	Windows 2000/NT	Windows 98/ 2000/NT	Windows 2000/XP/NT
Software								Catalyst
Costs								
Prototyper (RP system)	US-$	no longer available	33,500	67,500	175,000	49,995	29,900	Prototyper (RP system)

Table A2-14.2: 3D printing systems II

RP technology	Dimension	3D printing	3D printing		
Prototyper name / type		Objet Quadra	R 2	R 4	R 10
Manufacturer		Objet Geometries	Extrude Hone–ProMetal		
Equipment					
Dimensions	inch	51 / 47.2 / 35	72 / 48 / 60	72 / 68 / 84	108 / 84 / 71
Weight	lbs.	550	2100	4620	5000
Power supply	V / A	110	3 x 240 / 50	3 x 240 / 50	3 x 240 / 100
Power consumption (maximum)	kW	3.5			
Operating gas	cu. feet/min		compressed air (80-100 psi)		
Water supply	gal/min				
Extracting system capacity	cu. feet/min				
Chamber temperature	°F / °C	65 – 82 / 18–28			
Relative humidity	%	30–80			
Process					
Emissions (vapor, gas)					
Disposal / household garbage					
Dangerous waste					
Shaping element					
Technology		print head	print head	print head	print head
Type					
x-y-contouring		printer / plotter	printer / plotter	printer / plotter	printer / plotter
Outline accuracy	inch	600 / 300 dpi (x-y)	± .001	± .001	± .001
Repeat accuracy	inch				
z-outline accuracy	inch	1270 dpi			
z-outline repeat accuracy	inch				
Recoating cycle time	s				
Recoating velocity	inch/s				

Table A2-14.2 (cont.)

RP technology	Dimension	3D printing	3D printing		
Power fluctua-tion	%				
Part properties					
Maximum build volume	inch	10.6 x 11.8 x 7.8	8 x 8 x 6	15.7 x 15.7 x 10	40 x 20 x 10
Maximum part dimensions	inch	10.6 x 11.8 x 7.8	8 x 8 x 6	15.7 x 15.7 x 10	40 x 20 x 10
Layer thickness	inch				
Beam spot size	inch				
Accuracy	inch/inch				
Repeat accuracy	inch				
Support structure					
Interface					
Input data formats		STL	STL	STL	STL
Information processing system		PC	PC	PC	PC
Operating system		Windows NT	Windows 2000	Windows NT	Windows NT
Software			ProMetal Software	ProMetal Software	ProMetal Software
Costs					
Prototyper (RP system)	US-$	on demand	175,000	325,000	625,000

Table A2-15: Survey of photopolymers

Manu-facturer	Group of materials	Type		Properties	Laser / Lamp	www
		Acrylate	Epoxy			
Ciba–SL materials taken over by Ventico	Cibatool					www.vantico.com
	SL 5154	X		productive, quick	Ar$_+$	
	SL 5170		X	precise, transparent	HeCd	
	SL 5180		X	versatile	Ar$_+$	
	SL 5190		X	dimensionally stable, transparent	Solid State	
	SL 5195		X		Solid State	
	SL 5210	vinyle ether		water-resistant, heat-resistant	HeCd	
	SL 5220		X	moisture-resistant	HeCd	
	SL 5410		X	moisture-resistant	Ar$_+$	
	SL 5510		X	moisture-resistant	Solid State	
	SL 5520		X	impact resistant, elastic	Solid State	
	SL 5530		X	high-tempera-ture resistant	Solid State	
DuPont	Somos [1]					www.dupont.com
	2110	X		elastic	HeCd	
	2100	X		elastic	Ar$_+$	
	3110	X		elastic	HeCd	
	3100	X		elastic	Ar$_+$	
	6110		X	dimensionally stable, quick	HeCd	
	6100		X	dimensionally stable, quick	Ar$_+$	
	6120		X	dimensionally stable, quick	Solid State	
	7100		X	moisture- & heat-resistant	Ar$_+$	
	7110		X	moisture- & heat-resistant	HeCd	
	7120		X	moisture- & heat-resistant	Solid State	
	8100		X	impact resistant, elastic	Ar$_+$	
	8110		X	impact resistant, elastic	HeCd	
	8120		X	impact resistant, elastic	Solid State	

Table A2-15 (cont.)

Manu-facturer	Group of materials	Type		Properties	Laser / Lamp	www
		Acrylate	Epoxy			
RPC	RP Cure (2)					www.rpc.ch
	100 HC		X	very high r esolution, moisture-resistant	HeCd	
	100 AR		X		Ar_+	
	100 ND		X		Solid State	
	200 HC		X	impact resistant, elastic	HeCd	
	200 AR		X	impact resistant, elastic	Ar_+	
	300 HC		X	high-tempera-ture resistant	HeCd	
	300 AR		X	high-tempera-ture resistant	Ar_+	
	300 ND		X	high-tempera-ture resistant	Solid State	
	500 HC	X		heat-resistant, sterilizable	HeCd	
Solimer	Solimer (3)					out of business since 2001
	XA-7501	X			UV lamp	

(1): independent resin manufacturer, adapted to all available systems; preferred use in EOS- and F&S-systems
(2): adapted to all available systems
(3): exclusive supplier of Cubital systems

Table A2-16: Physical properties of SL resins

Properties	Dimension	SL 5154	SL 5170	SL 5180	SL 5190	SL 5195	SL 5210	SL 5220	SL 5410	SL 5510	SL 5520	SL 5530
Density (77°F / 25°C)	g/cm³	1.12	1.14	1.15	1.15	1.16	1.15	1.14	1.16	1.13	1.15	1.19
Viscosity (77°F / 25°C)	Pa s	2.997	1.8 (86°F)	2.65	2.5	2.3	3.05 (86°F)	2.75 (86°F)	5.6	1.8 (86°F)	4.4 (86°F)	3.1
Flash point	°F / °C	329 / 165	361 / 183	> 248 / > 120								
Thermal decomposition	°F / °C	320 / 160	320 / 160	302 / 160								
Ignition temperature	°F / °C	860 / 460	680 / 360	680 / 360								
Penetration depth	in	.0051	.0047	.0043	.0047	.0051	.0047	.0055	.0047	.0039	.0039	.0055
Critical exposure rate	mJ/cm²	4.2	13.5	13.3	17.7	13.1	5	9	10.1	11.2–11.4	13–19.9	8.9–9.4

Properties	Dimension	2110 / 2100	3110 / 3100	6100	6110	6120	7100	7110	7120	8100	8110	8120
Density (77°F / 25°C)	g/cm³	1.16	1.13	1.15	1.15	1.15	1.13	1.13	1.13	1.11	1.11	1.11
Viscosity (77°F / 25°C)	Pa s	3.5 / 4.5	11 / 9	3.9	3-4 (86°F)	3.9	8 (86°F)	8 (86°F)	8 (86°F)	6 (86°F)	6 (86°F)	6 (86°F)
Flash point	°F / °C											
Thermal decomposition	°F / °C											
Ignition temperature	°F / °C											
Penetration depth	in	.0055 / .0059	.0051 / .0074	.0047	.0055	.0071	.0055	.005	.0047	.0063	.0055	.0063
Critical exposure rate	mJ/cm²	3 / 2.5	2.5 / 4	12.2	8	17.5	10	8.2	8	7.2	6	6.75

Properties	Dimension	100 HC	100 AR	100 ND	200 HC	200 AR	300 HC	300 AR	300 ND	500 HC	500 HC	XA-7501
Density (77°F / 25°C)	g/cm³	1.14	1.14	1.14	1.15	1.15	1.15	1.15	1.15	1.12	1.12	1.1
Viscosity (77°F / 25°C)	Pa s	6.5 (86°F)	5.4 (86°F)	5 (86°F)	4.8 (86°F)	4.4 (86°F)	5.3 (86°F)	5 (86°F)	5.4 (86°F)	14.6 (86°F)	13.7 (86°F)	4
Flash point	°F / °C	356 / 180	356 / 180	356 / 180	> 302 / > 150	> 302 / > 150	> 302 / > 150	> 302 / > 150	> 302 / > 150	352 / 178	352 / 178	
Thermal decomposition	°F / °C	> 392 / > 200	> 392 / > 200	> 392 / > 200	> 392 / > 200	> 392 / > 200	> 392 / > 200	> 392 / > 200	> 392 / > 200	> 662 / 350	> 662 / 350	
Ignition temperature	°F / °C											
Penetration depth	in	.0043	.0047	.0047	.0047	.0047	.0051	.0047	.0047	.010	.0051	
Critical exposure rate	mJ/cm²	10	7.7	10.6	8.3	7.8	10.5	10.1	10.8	9	5.4	

Table A2-17: Mechanical properties of cured SL resins

Properties	Dimension	SL 5154	SL 5170	SL 5180	SL 5190	SL 5195	SL 5210	SL 5220	SL 5410	SL 5510	SL 5520	SL 5530
Tensile strength	MPa	35	59-60	46.5	55-57	46.5	15	62	72	77	26-33	56-61
Tensile modulus	MPa	1100-1200	2400-2500	2090	2150-2250	2090	1455	2703	3095	3296	1034-1379	2889-3144
Elongation at break	%	11-19	7-19	6-16	9	11-22	1.2	8.3	4.2	5.4	23-43	3.8-4.4
Flexural modulus	MPa	770	2920-3010	1628	2110-2450	1628	1724	2951	42848	3054	689-896	2620-3240
Flexural strength	MPa		107-108	49.3	75-90	49.3	44	94	127	99	29-41	63-87
Impact strength	kJ/m²	20-25	27-30	32-47	27	54	21	37	39	27	59	21
Shore hardness	Shore D	78	85	84	80	83	84	86	86	86	80	88

Properties	Dimension	2110 / 2100	3110 / 3100	6100	6110	6120	7100	7110	7120	8100	8110	8120
Tensile strength	MPa	7 / 9	21	54	69	43	62	56	58	26	18	26
Tensile modulus	MPa	34 / 80	810	2690	2800	2200	2250	2117	2477	276-738	186	276-704
Elongation at break	%	75.8 / 46.7	9.2	7-11	10	4	5	8.1	6.9	17-24	27	18-30
Flexural modulus	MPa	34 / 124	310	2050	2300	1900	2650	2434	2967	628	310	690
Flexural strength	MPa							85	108	26	11	26
Impact strength	kJ/m²	86 / 90	15	33.8	27	32	29	27.8	27	59	87	59
Shore hardness	Shore D	53 / 62	80	84.5	87	88	86	82	88	81	77	76

Properties	Dimension	100 HC	100 AR	100 ND	200 HC	200 AR	300 HC	300 AR	300 ND	500 HC	500 HC	XA-7501
Tensile strength	MPa	67	75	75	50	50	67	60	60	35	35	
Tensile modulus	MPa	3210	3020	3000	2000	2000	3000	3000	3000	1300	1300	700
Elongation at break	%	4	9	9	17	17	3	3.5	3.5	9	9	20
Flexural modulus	MPa	2500	2400	2400	1500	1500	2500	2500	2500	1000	1000	1300
Flexural strength	MPa											45
Impact strength	kJ/m²	21	30	30	no break	no break	7	15	12	18	18	18
Shore hardness	Shore D	85	83	82	82	80	86	85	84	80	80	80

Table A2-18: Thermal properties of cured SL resins

Properties	Dimension	SL 5154	SL 5170	SL 5180	SL 5190	SL 5195	SL 5210	SL 5220	SL 5410	SL 5510	SL 5520	SL 5530
Glass transition temperature	°F °C	181 83	149–194 65–90	149–185 65–85	140–176 60–80	– –	140 60	127 53	154–190 68–88	154 68	94 34	174 79
Temperature of dimensional stability	°F °C	– –	120–131 49–55	107–120 42–49	109–122 43–50	116 47	115 46–47	107 42	131–190 55–88	143 62	111 44	176 80
Coefficient of thermal expansion	ppm/K	151 / 158.2	90	104.4								
Thermal conductivity	W/m K	0.198	0.2002	0.2099								

Properties	Dimension	2110 / 2100	3110 / 3100	6100	6110	6120	7100	7110	7120	8100	8110	8120
Glass transition temperature	°F °C	109 43	109 43	122 50	122 50	122 50	187 86					
Temperature of dimensional stability	°F °C	– / 111 – / 44		156 69	154 68	156 69	215 102	138–161 59–72	< 158 < 70	129 54	111–120 44–49	140 60
Coefficient of thermal expansion	ppm/K											

Properties	Dimension	100 HC	100 AR	100 ND	200 HC	200 AR	300 HC	300 AR	300 ND	500 HC	500 HC	XA-7501
Glass transition temperature	°F °C	177 81	177 81	177 81	140 60	140 60	248 120	248 120	248 120	197 92	197 92	161 72
Temperature of dimensional stability	°F °C	194 90	194 90	194 90	149 65	149 65	302 150	302 150	302 150	212 100	212 100	
Coefficient of thermal expansion	ppm/K		63.7									

Table A2-19: Electrical properties of cured resins

Properties	Dimension	SL 5154	SL 5170	SL 5180	SL 5190	SL 5195	SL 5210	SL 5220	SL 5410	SL 5510	SL 5520
Relative dielectric constant	at										
	0.05 kHz	4.6	4.2	4.1							
	0.1 kHz	4.5	4.2	4.1							
	1 kHz	4.3	4.2	4.1							
	10 kHz	4.2	4.1	4							
	100 kHz	4	3.9	3.9							
	1000 kHz	3.7	3.8	3.8		For these materials no electrical properties are published.					
Dielectric loss factor	at										
	0.05 kHz	3.1	0.4	0.7							
	0.1 kHz	3	0.6	0.8							
	1 kHz	2.9	1.2	1.2							
	10 kHz	2.9	2	1.6							
	100 kHz	3.3	2	1.6							
	1000 kHz	3.9	2.4	2.2							
Specific volume resistance	W cm	9.8E13	1.7E15	1.44E15							
Specific surface resistance	W cm	8.7E13	2.4E12	5.9E13							
Electrical breakdown field	kV/mm	15.3	18.9	20.9							

Optical Tension Coefficient S at Room Temperature versus Load
for various completely Cured Stereolithography Resins

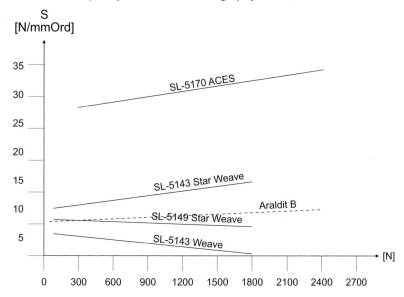

Optical Tension Coefficient S at Freezing Temperature versus Load
for various completely Cured Stereolithography Resins

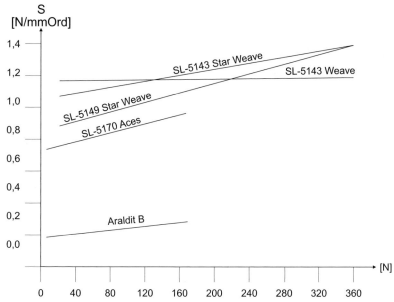

Figure A2-1 Optical properties of selected stereolithography resins

Table A2-20: Properties of sintering materials I

Properties	Dimension	Dura-Form PA	Dura-Form GF	Somos 201 TPE	True-Form PM	Cast-Form PS	Copper PA	Rapid Steel 2.0	Sand-Form SI	Sand-Form ZR 2.0
Type		polyamide	polyamide, glas filled	elastomer	polymer	wax infiltrated		bronze infiltrated		Circonium sand
Manufacturer / distributor		DTM	DTM	DuPont / DTM	DTM	DTM	DTM	DTM	DTM	DTM
Color		white	grey	old English white	grey	white	red	grey	yellow	yellow
Middle particle size	μm	58	48	93	33	62	48	34	99	97
Particle size	95 % < μm	92	92	190	60	106	106	58	~200	~200
Range of particle sizes	μm	25–92	10–92	23–190	15–60	25–106	17–106	20–58	53–200	60–200
Absorption of humidity	%	0.41	0.3		0.26					
Specific gravity	g/cm^3	1.01–1.02		42736						
Density (powder)	g/cm^3	0.59	0.84	0.58	0.3–0.4		1.8	8.15	2.5	4.7
Density (sintered mat.)	g/cm^3	0.97	1.4	0.91	1.08		3.45	7.5	1.6	
Elastic modulus	MPa	1600	5910		1100			263_10^3		
Breaking elongation	%	22	2	111	1.2			0.9		
Flexural modulus	MPa	1430	3300	16.3			3200			
Flexural strength	MPa						55			
Charpy impact strength	kJ/m^2	432	101		21.9					
Charpy notch impact strength	kJ/m^2	216	96		8.1					
Tensile strength	MPa	44	33	20	10		35	580	677	600
Hardness				SA 81			SD 75	HRC 22		
Residual ash	%				1.1	0.03				
Melting point	°F / °C	363 / 184	365 / 185	313 / 156			356 / 180		3110 / 1710	3810 –4170 / 2100 –230
Glass transition temperature	°F / °C				156 / 69	192 / 89				
Temperature of dimensional stability	°F / °C	350 / 177	347 / 175		143 / 62		318 / 176			

Table A2-21: Properties of sintering materials II

Properties	Dimension	Direct-Steel 50-V1	Direct-Metal 50-V2	Direct-Metal 100-V3	PA 1500	PA 1300 gf	PA 2200	PA 3200 gf	PS 2500	EOSINT S quartz	EOSINT S zircon HT
Type					polyamide	polyamide, glass filled	fine polyamide	fine polyamide, glass filled	polystyrene	quartz sand	zirconium sand
Manufacturer / distributor		Electrolux / EOS			EOS	EOS	EOS	EOS	EOS	EOS	EOS
Color		grey	brown	brown	white	white	white	white	white	brown	brown
Middle particle size	µm	50	50	100	110	80	50	50	85	160	190
Particle size	95 % < µm										
Range of particle sizes	µm										
Absorption of humidity	%										
Specific gravity	g/cm³	ca. 8.2	ca. 9	ca. 9	1.04		1.03				
Density (powder)	g/cm³	ca. 4.3	ca. 5.1	ca. 5.7	0.5	0.85	0.45–0.5	0.70–0.75	0.45	1.4	2.45
Density (sintered mat.)	g/cm³	ca. 7.8[1]	ca. 6.3[1]	ca. 6.6[1]	0.90–0.10[1]	1.3–1.40[1]	0.90–0.95[1]	1.2–1.3[1]	0.5–0.7[1]	< 1.4[1]	<2.45[1]
Elastic modulus	MPa				1000	1600	1500	2800[1]	550		
Breaking elongation	%				> 25	3	15	2.8–6.0[1]	1.5		
Flexural modulus	MPa										
Flexural strength	MPa	950[1]	300–400[1,2]	300[1]						800 N/cm² (cold, post-cured)	
Charpy impact strength	kJ/m²				without break		21	15	12.5		
Charpy notch impact strength	kJ/m²				21		3	2.7			
Tensile strength	MPa	500[1]	120–200[1,2]	180[1]	40	40	50	40–47	5		
Hardness	%				70[3]		90[3]				
Residual ash	%								0.05		
Melting point	°F / °C	> 1292 / > 700	> 1292 / > 700	> 1292 / > 700	ca. 356 / 180	ca. 356 / 180	ca. 356 / 180	ca. 356 / 180	amorphous		
Glass transition temperature	°F / °C								ca. 212 / 100		
Temperature of dimensional stability	°F / °C				284 / 140	284 / 140					

(1): dependent on process parameters (adjustable)
(2): without infiltration / resin-infiltrated
(3): ball-pressure hardness

Table A2-22: Properties of FDM materials

Properties	Dimension	ICW06	P400	P500	E20	P1500
Type		precision casting wax	ABS	ABSi	elastomer	polyester
Manufac-turer				Soligen		
Elastic modulus	MPa	280	2517	1993	68	839
Breaking elongation	%	<10	>10	>10	>10	<10
Flexural modulus	MPa	280	2657	1797	137	839
Flexural strength	MPa	4,3	66	59	5	28
Impact strength	kJ/m^2	49	1282	1068	377	1228
Notch impact strength	kJ/m^2	17	107	176	347	32
Tensile strength	MPa	3.6	35	38	6	9,6
Shore hardness	Shore D/A	33	78	76	40/96	62
Melting point	°F °C	147 64	464 200	464 200	264 129	302 150
Operating temperature	°F °C	154 68	518–554 270–290	518–554 270–290	320 160	347 175
Density	g/cm^3	1	1.05	1.07	1.32	1.1

Remarks:
ICW06: residual ash content <0.004%
P400:   available in: white, black, red, blue, green, yellow and in any color if defined in color codes
        (RAL, Pantone)
P500:   fulfills all FDA USP CLASS VI requirements, gamma-sterilizable

Table A2-23: Properties of ModelMaker materials

Properties	Dimension	Protobuild	Protosupport
Type		Green build material	Red support material
Manufacturer			
Density	g/cm³	1.25	0.93
Melting point	°F °C	197–223 92–106	122–162 50–72
Ignition temperature	°F °C	> 342 > 175	> 342 > 175
Shore hardness	Shore A	45	33
Coefficient of thermal expansion			
at 50 °C	ppm/ °C	82.7	351.3
at 55 °C	ppm/ °C		403.2
at 60 °C	ppm/ °C	87.6	
at 70 °C	ppm/ °C	91.4	

Table A2-24: Properties of LOM materials

Properties	Dimension	LPH 042		LXP 050		LGF 045	
Type		paper		polyester		fiberglass	
Manufacturer		Cubic Technologies					
Fiber orientation		longi-tudinal	trans-verse	longi-tudinal	trans-verse	longi-tudinal	trans-verse
Density	g/cm^3	1.449		1.0–1.3		1.3	
Elastic modulus	MPa	2524		3435			
Tensile strength	MPa	26	1.4	85		> 124.1	4.8
Compressive strength	MPa	15.1	115.3	17	52		
Compression modulus	MPa	2192.9	406.9	2460	1601		
Maximum deformation at pressure	%	1.01	40.4	3.58	2.52		
Flexural strength	MPa	2.8–4.8		4.3–9.7			
Glass transition temperature	°F / °C	86 / 30				127–260 / 53–127	
Coefficient of expansion	ppm/K	3.7	185.4	17.2	229	X: 3.9 Y: 15.5	Z: 111.1
Thermal conductivity	W/mK	0.117					
Ash content	%	2.7–3.1					

Table A2-25: Properties of selected vacuum casting materials

Properties	Dimension	PX 774	2170	Biresin G 55	MG 410	Megithan 2MD787
Manufacturer		Axson	HEK	SIKA	ebalta	BIOTOOL
Distinguished properties		elastic	solid	impact resistant, flexural resistant	high-temperature resistant	crystal-clear (UV resistant)
Aspect / color		black	light yellow	opaque	light yellow	transparent
Mixing ratio A:B	g:g	44 : 100	100 : 150	80 : 100	100 : 200	100 : 300
Pot life (g;°C)	min	2–3 (100; 25)	4 (100; 25)	4 (300; 20)	1.5 (-; -)	15–20 (200; 20)
Viscosity (25 °C) Component A	Pa s	0.025–0.045	0.63	1.45 (23 °C)	0.4–0.5	1.5
Component B	Pa s	0.8–1.2	0.23	0.3 (23 °C)	0.05–0.07	0.8–2.2
Specific gravity (77°F) Component A	kg/dm³	1.18–1.22	1.11	1.06 (23 °C)	1	
Component B	kg/dm³	1.03–1.07	1.22	1.22 (23 °C)	1.2	
Hardness	Shore A Shore D	75 (23 °C)	82 (23 °C)	81–84 (23 °C)	86 (25 °C)	85–87 (25 °C)
Flexural strength	MPa		120	110–120	60–70	
Flexural modulus	MPa		2690	2800		
Tensile strength	MPa	7	78.5	70–80	60–70	90
Tensile modulus	MPa		3530			
Impact strength	kJ/m²			>100		
Elongation at break	%	300	6.5	6–8	5–6	
Elongation to break	%		5.5			
Shrinkage	%		0.1	0.25–0.35	0.5–0.7	0.1
Thermal conductivity	W/m K		224			
Heat resistance	°F / °C	176 / 80	185 / 86	176 / 80	>392 / > 200	194–212 / 190–100

The table above shows only a few sample materials.
The user should contact the manufacturer to find out the best suitable material for his application.

A3 Abbreviations

ACES   Accurate Clear Epoxy Solids, build style, 3D Systems
AFM   Anatomic Facsimile Imodels
AIM   ACES Injection Molding
CIP   Cold Isostatic Pressure
CPDM   Product Data Management, CIMATRON
CT   Computed Tomography, Computer Tomography
DLP   Digital Light Processing (Hewlet Packard)
DMD   Digital Micromirror Device
DMU   Digital Mok Up
ERP   Enterprise Recource Planning
ERM   Enterprise Recource Management
FEM   Finite Element Method
GFV   Generative Fertigungsverfahren = Layer Manufacturing Technique (German)
HIP   Hot Isostatic Pressure
HSC   High Speed Cutting
HSPC   High Speed Precision Cutting (Kern Microtechnik)
LENS   Laser Engineered Net Shaping
LMP   Layer Milling Process
LMC   Layer Milling Center
MIM   Metall Injection Moulding
MRT   Magnet Resonance Tomography
NMR   Nuclear Magnetic Resonance
PDM   Product Data Management
RIM   Reaktion Injection Moulding
SLC   SLC "Sliced Layer Contour",
       Various mostly proprietary formulations, such as:
       SLC, SLI   (3D-Systems)
       CLI   (EOS)
       HPGL   (HP)
       SLC   (Stratasys)
       F&S   (Fockele und Schwarze)
SLI   Slice data format, see SLC
SFF   Solid Freeform Fabrication
SFM   Solid Freeform Manufacturing
TI   Taylored Implants
VR   Virtual Reality
LSM   Laser Surface Melting

References and Bibliography

References as mentioned in the book

[ALT94]   Altmann, O.:Kunststoffteile mit Simultaneous Engineering kostenbewußt entwickeln. Kunststoffe 84 (1994) 12, p. 1728–1736.

[ARG91]   Argawalw, M. K., Bourell, D. L., Wu, B. et al.: An Evaluation on the Mechanical Behaviour of Bronze-Ni Composites Produced by Selective Laser Sintering. Solid Freeform Fabrication Symposium, Austin, Texas 1991.

[AUB94]   Aubin, R. F.: A World Wide Assessment of Rapid Prototyping Technologies. Beitrag zur IMS International Conference on Rapid Product Development, Stuttgart 1994.

[BEI94]   Beitz, W., Lamm, A., Ratfisch, U., et al.: Eine Systemumgebung zur Unterstützung von Simultaneous Engineering. VDI-Berichte 1148, p. 41–47, VDI Verlag, Düsseldorf, 1994.

[BER98]   Bernard, A., Taillander, G.: Le prototypage rapide. Editions HERMES, Paris 1998.

[BRE95]   Breitinger, F.: Zeitorientierte Strategien zur Herstellung von Prototypen-Spritzgußwerkzeugen. Diplomarbeit, Technische Universität München 1995.

[BUL95]   Bullinger, H.-J.: What makes Product Development Rapid? Proceeding of the International Conference on RPD, Rapid Product Development. Stuttgarter Messe- und Kongressgesellschaft mbH, Stuttgart 1995.

[BUL97]   Bullinger, H.-J., Hase, B.: Kundenorientiertes Qualitätsmanagement im FuE-Bereich. Deutschlandweite Unternehmensstudie, FhG-IAO, Stuttgart 1997.

[BUR93]   Burns, M.: Automated Fabrication – Improving Productivity in Manufacturing. PTR Prentice Hall, Englewood Cliffs, New Jersey 1993.

[CAR91]   Carter, W. T. Jr., Jones, M. G.: Direct Laser Sintering of Metals. Contribution to: Solid Freeform Fabrication Symposium, Austin, Texas 1991.

[DAA98]   Daas, M.: Entwicklungsbegleitendes Rapid Prototyping. Tagungsband der Konferenz: Von der Virtual Reality zum Rapid Prototyping, 10.–11. 11. 98 in Düsseldorf. Euroforum, Düsseldorf 1998.

[DAH98]   Dahmen, R.: Charakteristiken und Bewertung der (laserunterstützten) Rapid Prototyping Verfahren, insbesondere der LLM-Verfahren.
           Written final examination, University of Applied Sciences, Aachen, Germany, 1998.

[DIC95]   Dickens, Ph.: Investment Casting of Rapid Prototype Models. EARP Letter 7/1995, c/o Bent Mieritz, DTI, Aarhus, Denmark, 1995.

[DIL84]   Dillon, K., Terchek, R.: Dimensionally Controlled Cobald-Containing Preci-
           sion Molded Metal Article. US-Patent 4,431,449, 14. 02. 1984. 3M Corp.

[DON91]   Donahue, R. J., Turner, R. S.: Proceedings of the Second International Confer-
           ence on Rapid Prototyping. University of Daytona, 1991, p. 221.

[EOS98]   Press Release Euromold fair'98. EOS, Munich, 1998.

[FAS94]   Faßbender, M.: Ermittlung von prozeß- und bauteilspezifischen Parametern für
           die Prototypenfertigung von Turbinenteilen auf einer Stereolithographieanlage.
           Diplomarbeit, Rheinisch-Westfälische Technische Hochschule Aachen, 1994.

[FOK94]   Fockele, M., Schwarze, D.: Neue Wege im Rapid Prototyping. Beitrag zur 27th
           ISATA, Aachen 1994.

[GAR98]   Gartzen, J., Gebhardt, A. et al.: Optimisation using THESA – Thermo Elastic
           Tension Analysis is now beeing used for LASER-Sintered Components. Proto-
           typing Technology International 1998. UK & International Press, Dorking,
           Surrey, UK, 1998.

[GEB95]   Gebhardt, A.: Durch dick und dünn. Laser in der Materialbearbeitung.
           WT – Produktion und Management 85 (1995) 6, p. 314–317.

[GEB97]   Gebhardt, A., Mathar, W.: Quality Parts on a small Scale. Prototyping Technol-
           ogy International, p. 280 ff., UK & International Press, Dorking, Surrey, UK,
           1997.

[GIB02]   Gibson, Ian (Ed.), Software Solutions for Rapid Prototyping. Professional
           Engineering Publishing, Suffork, UK, 2002.

[GRI02]   Grimm, Todd. Stereolithography, Selective Laser Sintering and PolyJet: Evalu-
           ating and Applying the Right Technology. www.atirapid.com, 2002.

[GRU93]   Gruenwald, Geza Plastics: How Structure Determines Properties. Hanser,
           Munich, 1993.

[HEL95]   Heller R.: Experimental Determination of the Properties of the Photosensitive
           Resin „HS 660". Personal Information, 1995.

[JAC92]   Jacobs, P. F.: Rapid Prototyping & Manufacturing. Society of Manufacturing
           Engineers, Dearborn 1992.

[JAC94]   Jacobs, P. F., Partaman, J., Bedal, B.: Oberflächenqualität. The Edge, Vol. 3,
           No. 3, 1994, p. 6.

[JAC97]   Jacobs, P.F.: Recent Advances in Rapid Tooling from Stereolithography. 3D-
           Systems, published 3D-Systems Company Paper, 1997.

[JTE97]   Panel Report on Rapid Prototyping in Europe and Japan, Vol. 1. Analytic
           Chapters, March 1997. Loyola College in Maryland, Maryland, USA, 1997.

[KAR98]   Karjalainen, J., Tuomi, J. et al.: Quantifying the savings in product develop-
           ment. 7èmes assises du prototypage rapide, 19.–20. 11. 98, Paris 1998.

[KRE98]   Kreicher D. M., Miller W. D.: Laser Engineered Net Shaping (LENS): Beyond Rapid Prototyping to direct fabrication. 7èmes assises du prototypage rapide, 19.–20. 11. 98, Paris 1998.

[KRU91]   Kruth, J. P.: Material Incress Manufacturing by Rapid Prototyping Techniques. Annuals of the Cirp, Vol. 40/2, 1991.

[KUN97]   Kunze, H.-D.: Competitive Advantages by Near-Net-Shape Manufacturing. DGM Informationsgesellschaft Verlag, Frankfurt 1997.

[LEH94]   Lehmann, O., Stuke, M.: Three-dimensional laser direct writing of electrically conducting and isolating microstructures. Materials Letters 21, Okt. 1994, p. 131–136.

[MAI]   Maihack, St.: Schnittstellentools für neutrale CAD-Schnittstellen. CAD-CAM-Report 13 (1994) 9.

[MAR91]   Mariques-Frayre, J. A., Bourell, D. L.: Selective Laser Sintering of Cu-Pb/Sn Solder Powders. Solid Freeform Fabrication Symposium, Austin, Texas 1991.

[MIG96]   Migliore, Leonard, (Ed): Laser Materials Processing. Marcel Dekker Inc, New York 1996.

[NAC98]   Nachtrodt, M.: Rapid-Prototyping mit SOM. Beitrag zur Tagung: Die Wirtschaft im Dialog mit der Forschung, 25.–27. 11. 98. Forum Niederberg, Velbert 1998.

[NOE97]   Nöken, S.: Technologie des Selektiven Lasersinterns von Thermoplasten. In: Berichte aus der Produktionstechnik, Band 8/97. Shaker Verlag, Aachen 1997.

[RAN97]   Ranky, Paul G. An Introduction to Concurrent / Simultaneous Engineering, Methods, Tools and Case Studies (Integrated Product and Process Design). Book and CD Set edition, CIMware USA, Inc.; October 15, 1997.

[REI83]   Reinertsen, D. G.: Whodunit? The Search for the New-Product Killers. Electronic Business July 1983, p. 62–66.

[SCHU96]   Schulz, H.: Hochgeschwindigkeitsbearbeitung – High-Speed Machining. Carl-Hanser Verlag, München, Wien 1996.

[SER95]   Serbin, J., Wilkening, Ch., Bretsch, Ch. et al.: STEREOS and EOSINT. The New Developments and State of the Art. Produktinformation der EOS GmbH, München, 1995.

[SHE94]   Shellabear, M.: Rapid Prototyping: Verfahren und Materialien für Fein- und Sandgießverfahren. Gießerei 81 (22) 1994.

[SIE91]   Siegwart, H., Sieger, U.: Neues Verfahren für die Wirtschaftlichkeitsbeurteilung von Investitionen in neue Produktionstechnologien. Kostenrechnungspraxis 2 (1991), p. 63–70.

[STA01]   Stahlschlüssel , "Key to Steel", (Handbook of Steel), 19th edition 2001. Verlag Stahlschlüssel Wegst, Germany, 2002. www.stahlschluessel.de

[STE95]   Steger, W., Conrad, T.: Rapid Prototyping. Operative und strategische Bewertung von generativen und konventionellen Fertigungsverfahren. VDI-Z Special Werkzeug- und Formenbau, November 1995, p. 12 ff., VDI Verlag, Düsseldorf, 1995.

[STE94]   Steinchen, W., Kupfer, G., Kramer, B.: Photoelastic Investigations by means of Stereolithography. Beitrag zur 27th ISATA, Aachen, 1994.

[VAN95]   Van Crüchten, M.: Vergleich der generativen Fertigungsverfahren Stereolithographie und selektives Lasersintern am Beispiel der Modellierung eines Baggerarms. Diplomarbeit, Rheinisch-Westfälische Technische Hochschule Aachen, 1995.

[VAR01]   Varadan, Vijay, K.: Microstereolithography and Other Fabrication Techniques for 3D MEMS. John Wiley & Sons New York, USA, 2001.

[VDI1]   VDI-Report 1148: VDI-Verlag, Düsseldorf, 1993.

[VDI22]   VDI-Richtlinie 2221: Methodik zum Entwickeln und Konstruieren technischer Systeme und Produkte. (Methodical Approach to the Development and Design of Technical Systems and Products), VDI-Verlag, Düsseldorf 1993.

[WOH98]   Wohlers, T.: Worldwide Developments and Trends in Rapid Prototyping and Tooling. Tagungsband der ICRPM '98, International Conference on Rapid Prototyping, Manufacturing and Rapid Tooling, 21.–23. 07. 1998 in Beijing. Shaanxi Science and Technology Press, Xian, Shaanxi, PRChina, 1998.

[WOH99]   Wohlers, T.: Rapid Prototyping & Tooling, State of the Industry: 1999, Worldwide Progress Report. Wohlers Associates Inc., Fort Collins, CO, USA 1999.

[WIL95]   Wildung, D.: Nefertitit's Mother-in-Law – Rapid Prototyping and Ancient Egypt. International Conference on RPD Rapid Product Development. Stuttgarter Messe- und Kongressgesellschaft mbH, Stuttgart, 1995.

[WIL95]   Willinger, H.-M.: Kostenrechnung Rapid-Prototyping Verfahren. Diplomarbeit, Technische Universität München, 1995.

Bibliography

Baraldi, U.:   RAPITOOL: Rapid and Economical Production Techniques for Prototypes and Short Runs Thermoplastic Injection Moulds. Proceedings of the 27th ISATA, Aachen, 1994.

Kochan, D.:   Solid Freeform Manufacturing – Advanced Rapid Prototyping. Elsevier Science Publishers B. V., Amsterdam, London, New York, Tokyo, 1993.

Naber, H.:   Fast Prototype Tools. In: Rapid Prototyping and Manufacturing '95.
Macht, M.,   Society of Manufacturing Engineers, Dearborn, 1995.
Geuer, A.

Schueren,   Laser Based Selective Metal Powder Sintering: a Feasibility Study.
van der B.,   LANE '94, Erlangen, 1994.
Kruth, J. P.:

Spencer, J. D.:   Surface Finishing Techniques for Rapid Prototyping. Rapid Prototyping
                     Conference, Dearborn, Michigan, 1993.

Wyatt, M., :   Choosing a Solvent to clean Stereolithography Parts. Prototyping
Dishart, K.:   International 98. P. 107–109. U.K. & International Press, Dorking, UK,
                     1998.

McDonald, J. A.   Rapid Prototyping Casebook. Professional Engineering Publishing Ltd.,
et al.:   London and Bury St. Edmunds, UK, 2001.

Index

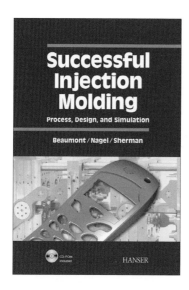

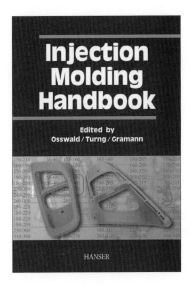

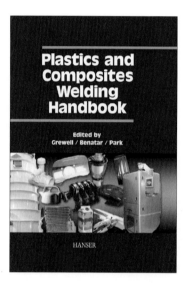

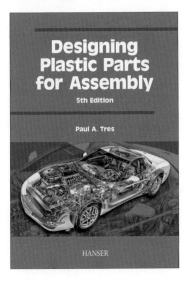